PENGUIN BOOKS

THE DIFFERENCE ENGINE

Doron Swade is Assistant Director and Head of Collections at the Science Museum in London. He is an engineer and a historian of technology, and a leading authority on the life and work of Charles Babbage. He has published widely on curatorship and the history of computing and appears frequently on radio and television.

Doron Swade masterminded the project to construct a Babbage calculating engine from original nineteenth-century designs. The vast engine was completed in 2002 after seventeen years of engineering, plotting, and politics. He was born in Cape Town, South Africa, and lives in London, England.

Praise for *The Difference Engine*

Chosen as a Book of the Year by the *Economist* and as an Amazon.com Editors' Best of 2001 in the Digital Business and Culture Category

"An unstintingly honest portrayal of the man and the genius."
—*The Dallas Morning News*

"You'll want to tell your friends how an Englishman named Charles Babbage dreamed up a machine that could be programmed to perform complicated mathematical functions and print the results—years before the vacuum tube or even the electric light."
—*Houston Chronicle*

"Fun . . . the kind of gee-whiz book you can't help but read out loud to your friends." —*Chicago Tribune*

"Love, money, untimely death, politics, nobility, venality and a passion for mathematics and technology—what more can a science reader ask for?" —*Pittsburgh Post-Gazette*

"*The Difference Engine* is lovingly comprehensive and will thrill readers."
—Amazon.com

"Swade brings a new, deep-tech perspective as the director of a recent six-year effort that built a working engine from Babbage's original blueprints." —*Wired*

"In this worthy volume, Swade has given us a human portrait of a flawed visionary and, in the bringing of the Babbage Engine to physical existence, a tale of modern times."

—*San Jose Mercury News*

"*The Difference Engine* is in every way a worthy successor to the mantle of Dava Sobel's *Longitude*."

—*The Oregonian* (Portland)

"An absorbing, readable account of Babbage's obsessive quest . . . fascinating." —*The Seattle Times*

"Any new biographer would face a challenge in finding original material, but Swade amply answers the call."

—*Booklist*

"Babbage's eccentricities and his struggles, both personal and professional, are deftly sketched."

—*South Bend Tribune*

"Swade shines when it comes to explaining what the world was like in Babbage's time and the challenges that Babbage faced in creating the Engine." —*Technology and Society* (TechSoc.com)

"A wonderful read." —Martin Campbell-Kelly, University of Warwick

"Engaging . . . Swade's able account of this gifted scientist, his cohorts and their curious endeavors enhances and broadens the growing body of literature on computer history."

—*Publishers Weekly*

"One of the most knowledgeable Babbage experts here tells two engaging stories for a popular audience: how Charles Babbage failed to build his engine, and how Doron Swade succeeded. Although separated by more than a century, both are fascinating tales replete with quirky personalities, engineering challenges, and edge-of-the-seat funding crises."

—Len Shustek, Chairman of the Computer Museum History Center

"Drama and humanity . . . fun . . . compelling . . . deserves the same success as *Longitude*." —*The Spectator* (London)

"As gripping as any thriller."

—*The Daily Telegraph* (London)

"A moving and fascinating account of a brilliant man."

—*Kirkus Reviews*

THE DIFFERENCE ENGINE

Charles Babbage and the Quest to Build the First Computer

DORON SWADE

PENGUIN BOOKS

PENGUIN BOOKS
Published by the Penguin Group
Penguin Putnam Inc., 375 Hudson Street,
New York, New York 10014, U.S.A.
Penguin Books Ltd, 80 Strand, London WC2R 0RL, England
Penguin Books Australia Ltd, 250 Camberwell Road,
Camberwell, Victoria 3124, Australia
Penguin Books Canada Ltd, 10 Alcorn Avenue,
Toronto, Ontario, Canada M4V 3B2
Penguin Books India (P) Ltd, 11 Community Centre,
Panchsheel Park, New Delhi – 110 017, India
Penguin Books (N.Z.) Ltd, Cnr Rosedale and Airborne Roads,
Albany, Auckland, New Zealand
Penguin Books (South Africa) (Pty) Ltd, 24 Sturdee Avenue,
Rosebank, Johannesburg 2196, South Africa

Penguin Books Ltd, Registered Offices:
Harmondsworth, Middlesex, England

First published in Great Britain by Little, Brown and Company
as *The Cogwheel Brain* 2000
First published in the United States of America by Viking Penguin,
a member of Penguin Putnam Inc. 2001
Published in Penguin Books 2002

1 3 5 7 9 10 8 6 4 2

THE LIBRARY OF CONGRESS HAS CATALOGED
THE AMERICAN HARDCOVER EDITION AS FOLLOWS:
Swade, Doron.
The difference engine : Charles Babbage and the quest to
build the first computer / Doron Swade.
p. cm.
ISBN 0-670-91020-1 (hc.)
ISBN 0 14 20.0144 9 (pbk.)
1. Calculators—Great Britain—History—19th century.
2. Babbage, Charles, 1791–1871. I. Babbage, Charles, 1791–1871.
QA75 .S954 2001
681'.145—dc21 00-068643

Printed in the United States of America
Set in Goudy

CONTENTS

Acknowledgements vii

A Note on the Value of Money in the Nineteenth Century x

Preface 1

Part I THE DIFFERENCE ENGINE

Chapter 1	**The Tables Crisis**	9
Chapter 2	**A Personal Question**	32
Chapter 3	**Tragedy and Decline**	49
Chapter 4	**Miracles and Machines**	72

Part II THE ANALYTICAL ENGINE

Chapter 5 **Breakthrough** 91
Chapter 6 **Applause in Turin** 114
Chapter 7 **The Astronomer Royal Objects** 134
Chapter 8 **The Enchantress of Number** 155
Chapter 9 **Intrigues of Science** 172
Chapter 10 **Visionaries and Pragmatists** 193
Chapter 11 **Curtain Call** 210

Part III A MODERN SEQUEL

Chapter 12 **Here We Go Again** 221
Chapter 13 **The Trial Piece** 232
Chapter 14 **The Money** 252
Chapter 15 **The Deal** 269
Chapter 16 **The Build** 283
Chapter 17 **The 'Irascible Genius' Redeemed** 296
Chapter 18 **The Modern Legacy** 308

Charles Babbage: Biographical Note 319
Bibliography 321
Credits 332
Index 333

ACKNOWLEDGEMENTS

There is a large and growing literature on Charles Babbage and his engines. This book draws freely on existing published sources as well as new archival material. While nothing in these pages is fiction, this book is not intended as an academic treatise and there are no running footnotes or endnotes to identify specific sources for quoted material. I wish therefore to acknowledge several particular debts. First to Bruce Collier's splendid doctoral work, *The Little Engines that Could've: The Calculating Machines of Charles Babbage*. This study of Babbage and his engines remains unsurpassed. To Anthony Hyman's *Charles Babbage: Pioneer of the Computer*, which contains the richest account of Babbage's life and the context of his times. To Allan Bromley's published and unpublished work on the technical design and workings of Babbage's calculating engines, which remains unequalled. To Michael

Lindgren's *Glory and Failure: The Difference Engines of Johann Müller, Charles Babbage and Georg and Edvard Scheutz*, which provides the most detailed exploration of the Swedish machines. To Betty Toole's *Ada, the Enchantress of Numbers*, which contains selected transcriptions of Ada Lovelace's correspondence.

Manuscript archives include the comprehensive set of Babbage's notebooks and design drawings held in the Science Museum Library in South Kensington; the volumes of Babbage's correspondence at the British Library; the General Register Office papers at the Public Record Office, Kew; the Greenwich Royal Observatory papers at the University of Cambridge; the Herschel papers at the Royal Society; and the Babbage archive held by Waseda University, Tokyo. To the archivists and librarians of these institutions I wish to express my gratitude and thanks for their patient help over the years. To C. J. D. Roberts for his limitless generosity with archival sources.

To Richard Keeler for his help on Babbage's ophthalmoscope. To Adam Perkins at the Cambridge University Library for his unfailingly helpful efforts on the Royal Observatory archives. To Dr Neville F. Babbage for his gracious encouragement and generous access to family material. To Martin Daunton for his wise guidance and optimistic efforts to transform an engineer into a historian. To Arthur Rowles for taking a risk on an ex-colonial ragamuffin twenty-five years ago, and for his friendship ever since.

This book would not have happened but for James Essinger. He originated the idea for a popular book on Charles Babbage and is responsible for its original title, *The Cogwheel Brain*. Without his boundless energy, unflagging commitment and promotional abilities the project would not have happened.

To colleagues and friends at the Science Museum from

whom I have learned so much. To Rod Smith, my collections assistant, for shielding me while I burrowed away in archives. To the Science Museum and the generations of curators and archivists who cared for and preserved the physical artefacts and documents on which this work is based, and for creating a supportive context in which such material can be studied.

To Sarah, who knows more about Babbage than she ever wanted to and who read and commented on the drafts.

A NOTE ON THE VALUE OF MONEY IN THE NINETEENTH CENTURY

Sums of money quoted in this book are given in contemporary terms. A gentleman in 1814, for example, could expect to support his wife and a few children in modest comfort on an income of £300 a year.

In 1971 Britain decimalised its currency and adopted pounds (£) and new pence (p). The currency in the nineteenth century was pounds (£), shillings (s) and pence (d). There were 20 shillings to the pound, and 12 pence to the shilling. A guinea was 21 shillings.

No attempt is made in the narrative to estimate the modern equivalent of the sums of money cited. There is no completely reliable way of translating the value of money from one age to another, and simple multipliers are often misleading. As a rough guide, the sums mentioned in Parts I and II of this text should be multiplied by between 30 and 150 times to give an indication of present-day values.

PREFACE

This book is a tale of two quests. The first is Charles Babbage's quest to realise a vision – that the science of number could be mastered by mechanism. By simply turning the handle of his massive calculating engine Babbage planned to achieve results which up to that point in history could be achieved only by mental effort – thinking. But this was not all. Calculating engines offered a tantalising new prospect. The 'unerring certainty' of mechanism would eliminate the risk of human error to which numerical calculation was so frustratingly prone. Infallible machines would compensate for the frailties of the human mind and extend its powers.

This tale unfolds in the most inventive decades of the nineteenth century, in which science and engineering flourished. It is a tale of squabbles over money, personal tragedy, a vendetta, a beautiful countess, a confrontation with the Prime Minister, political instability and public protests.

During the course of his struggle Babbage was led from mechanised arithmetic to the entirely new realm of automatic computing – a strange new land in which he was the first inhabitant.

The second quest is the twentieth-century sequel: the quest at the Science Museum to build a working Babbage engine in time for the bicentenary of Babbage's birth. This part of the tale is set in Thatcher's Britain and starts in the mid-1980s. Margaret Thatcher, the 'Iron Lady', Prime Minister since 1979, had catapulted Britain from the comfort of the welfare state into the unforgiving values of the market place. The Science Museum, part of the public sector, was not immune. 'Visitors' became customers. The book-lined Fellows Room, a quiet sanctuary that was once the Museum's library, became a prestigious venue for corporate hire.

The project to construct a working Babbage engine was one I had the good fortune to lead. The venture grew out of curatorial life at the Museum and the constant puzzling over the relationship between objects, history and meaning. It is a story of funding crises, technical setbacks, impossible deadlines, a company going bust and institutional politics. Along the way there were countless reasons why we might not succeed. Had we learned enough from Babbage? Could we prevail where Babbage had failed?

Charles Babbage came into my life in May 1985 when I was appointed curator of computing at the Science Museum in London. At the time I was an electronics engineer on the Museum's staff, designing interactive computer-based displays for the galleries which occupy some seven acres of public exhibition space. Engineers and scientists are trained largely without the civilising influences of history or philosophy, and I was no exception. The two years I spent studying philosophy of science and metaphysics at Cambridge in the early 1970s was a rewarding counterbalance, though it brought me no closer to the nineteenth century, to Babbage or to his work. At the time of my appointment I had heard of Babbage and his calculating engines but knew practically nothing of his life or of the history of his efforts.

As the new custodian of the computing collection I began to familiarise myself with the objects in my care. On public display and in hidden stores I found carefully inventoried collections of tokens, counters, pebbles, knotted cords and tally sticks, the earliest devices used for counting and arithmetical reckoning. I came across crafted instruments with graduated scales: sectors, quadrants and slide rules. There were the ornate desktop calculators of Blaise Pascal and his successors, intellectual giants of their time who aimed to relieve the drudgery of numerical calculation through mechanism.

I found mechanical desktop calculators – brass arithmometers with inscribed sliders and dials long since stilled – and comptometers with flat toadstool keys like those on a typewriter. There were card punches, automatic sorters and tabulators, the legacy of a revolutionary technology on which the great fortunes of IBM were founded in the 1920s and 1930s. There were examples of early electronic vacuum tube computers from the 1950s: room-sized, power-guzzling systems as effective as winter heating as they were at computation. And finally there were the electronic pocket calculators and personal computers that had poured out of factories in such alarming numbers and disconcerting variety.

But there was an incomparable prize which stood apart from everything else. This was the largest collection of physical relics of Babbage's efforts to construct his vast and intricate machines. This collection of trophies, all on public display, includes the experimental assembly of the Analytical Engine that was under construction at the time of Babbage's death, all he ever built of that revolutionary machine. Its modest size gives little clue to the monumental intellectual accomplishment of its conception and its much publicised role as the symbolic antecedent of the modern computer.

It was not long before I made my first acquaintance with what is perhaps the most celebrated icon in the prehistory of computing, all that Babbage built of his Difference Engine No. 1 – the 'finished portion of the unfinished engine' – and a poignant symbol of Babbage's unfulfilled ambitions. The 'portion' was completed by Babbage's engineer, Joseph Clement, late in 1832. True to its inventor's promise, it calculates without error. It works impeccably to this day and, in defiance of Babbage's detractors, provides compelling evidence for the feasibility of his early designs.

In a display case nearby I found the Swedish difference engine designed by Georg and Edvard Scheutz, father and son, and built by Bryan Donkin in London. This handsome engine, inspired by Babbage, was commissioned in 1857 by the Treasury for use in the General Register Office. Unlike Babbage's engines, the Scheutz difference engine has the distinction of having been finished. It stands in the company of Babbage's unfinished machines, a proud rival and a humble courtier.

There was more. I came across the section of what Babbage termed the Mill of the Analytical Engine built by his son Henry in 1910, and one of the six small demonstration pieces he assembled from unused parts. As well as the completed and partially completed portions of Babbage's engines, there were the remnants of his experimental pieces – small trial assemblies he constructed to test the principle and function of various mechanisms, sheet metal stampings, oddments of moulds and castings – the debris of his search for production techniques to make near-identical parts without the need for machining.

These experimental assemblies and the relics from Babbage's workshop were not the only bounty held by the Science Museum. The Museum's library holds a comprehensive technical archive of Babbage's work. It consists of over 500 large detailed design drawings as well as twenty volumes of Babbage's Scribbling Books – the manuscript journals which chronicle his thinking, often in cryptic unexplained sentences and scrawled diagrams.

It soon became clear that the Babbage material was the most important subset of objects in the computer collection, and I set about finding out more. I began browsing through the departmental books that I had inherited with the job. Almost every published history of computing had a section or chapter on Babbage's work. The bare bones of his saga were immediately clear. Babbage was the great pioneer of computing and was equally famous on two counts – for inventing computers and for failing to build them. In the Babbage story, inventive genius and failure are inseparable.

Almost all these accounts gave the same reason for Babbage's lack of success: the limitations of nineteenth-century engineering. The thesis, often stated and sometimes implied, is that parts could not be made precisely enough for the assembled machine to work. I had no reason at that time to doubt this assertion and accepted it at face value – after all, virtually every chronicle of Babbage's sorry saga cited it. The accounts, recycled from book to book, became predictable. Each time I came to the 'limitation of technology' thesis I flipped ahead, looking for the happy outcome. I assumed that as soon as mechanical engineering had evolved sufficiently for precision not to be a constraint, someone would have built a

Babbage engine. I was astonished to find that the saga of Babbage's failures had no sequel and unable to understand why so many chroniclers had been content to leave the tale unfinished. Why had no one attempted to build an engine, either as a tribute to Babbage or to test the most cited reason for his failures?

The circumstances surrounding the collapse of Babbage's own efforts to build his great engine were complex. There was an unresolved dispute with his engineer, funding crises, personality clashes and loss of credibility. Was it possible that these complex circumstances masked the technical or logical impossibility of Babbage's great invention? A tantalising question began to gnaw. *Could* Babbage have built his engine? And if so, would it have worked? This puzzle was the start of an adventure that dominated my life for the next six years, and to some extent continues to do so.

I have often imagined what Babbage's reaction might have been had he walked in on us finally assembling one of his machines. I visualise him in his long dress coat in the gallery having dropped in en route to some appointment, perhaps at the Royal Society, impatient to be on his way. He stands a short distance from us and takes in the scene at a glance. He does not speak. He nods and gives a grumpy, knowing smile.

Doron Swade
London, October 1999

Part I

The Difference Engine

Chapter 1

THE TABLES CRISIS

I wish to God these calculations had been executed by steam.

Charles Babbage, 1821

A carriage clatters to a halt outside No. 5 Devonshire Street, London. A well-dressed man in his late twenties alights with a bundle of manuscripts tucked under his arm. At the sound of the carriage, a tall gentleman comes out to greet him. The visitor is John Herschel, an astronomer, up in London from his father's house in Slough. His host is Charles Babbage, mathematician. It is summer 1821.

The two friends are pleased to see each other again. Their friendship had flowered during their undergraduate years at Cambridge, and they have stayed in close touch since. They exchange pleasantries as they go inside and swap some scientific gossip. Within a few minutes they sit down to their task in the drawing room.

Herschel divides the bundle of manuscripts equally between them. The sheets are covered with closely written calculations and lists of numbers – mathematical tables they are preparing for the Astronomical Society. The results in front of them have been calculated by 'computers'. These are not machines, but people who perform routine calculations by hand according to a fixed arithmetical procedure. The two stacks of papers contain the results of the same set of calculations carried out by different computers. If the computers have done their work perfectly, the two sets of results should be identical.

They settle in their chairs and start comparing the figures. Herschel reads a number from his sheet; Babbage checks it against the entry in front of him. Line by line they proceed in this painstaking way. They find an error – the result in Herschel's manuscript differs from Babbage's. The computers have made a mistake. The two friends mark the entry and proceed. Concentration is intense. At times they lose track, and have to go back. More errors. Babbage becomes increasingly agitated. Finally, he can contain himself no longer. 'I wish to God these calculations had been executed by steam.'

This exasperated appeal was made when Babbage was twenty-nine. Its ramifications were to dominate the rest of his life.

The burdens of calculation had plagued many before him, and Babbage was voicing the frustrations of centuries. Relief from the tyranny of numbers was sought in early mechanical aids to counting – tally sticks, pebbles and tokens – which date back several thousand years. There was a large gap until the shift from counting to mechanical calculation seized the imagination of some of the leading intellects of seventeenth-century Continental Europe. Wilhelm Schickard, Blaise Pascal and Gottfried Leibniz devised and constructed mechanical calculators to relieve the drudgery of doing sums. Leibniz's 'reckoner' built in the late 1670s used a revolutionary device, the stepped drum, which dominated calculator design for the next two centuries. The calculators of Leibniz and Pascal were sensations. They were paraded in the palaces of royalty and adorned the drawing rooms of the aristocracy, savants and the wealthy. But they were more in the nature of ornamental curiosities – *objets de salon* – exquisite, delicate, unreliable and unsuited to daily use.

By 1821, when Babbage and Herschel sat down to check their manuscripts, the situation was not much improved. Thomas de Colmar had just introduced a new calculating device, the arithmometer. This was a small desktop instrument with dials, sliders and a handle which, through a series of manual operations, was capable of basic arithmetic. The early arithmometers were erratic, and it was decades before they made their mark as practical devices for routine use. Slide rules were a great boon for quick and convenient calculation, but the scales and divisions were read by eye and there was an element of subjective judgement in the last decimal places. Accuracy was typically limited to three or four figures, which was fine for some purposes, but not all.

It was not until the last decades of the nineteenth century that mechanical desktop calculators became reliable, robust and cheap enough for general arithmetic – addition, subtraction, multiplication and division. Until then, scientists, engineers, surveyors and architects relied on printed tables for mathematical functions or calculations requiring more than a few figures of accuracy. Journeymen, builders, tradesmen, merchants and excise officers turned to printed books of look-up lists and 'ready reckoners' for multiplication, multiples of fractions, conversion of units and a host of arithmetical tasks essential to their daily work. Bankers, investors, actuaries, moneylenders and clerks relied on tables of interest for returns on investment, annuities and assurance premiums.

The bookshelves in studies, offices, ship's cabins and workshops groaned under the weight of volumes of tables for desk use, and well-thumbed pocket editions populated the bags and pouches of those on the move.

The need for tables and the reliance placed on them became especially acute during the first half of the nineteenth century, which witnessed a ferment of scientific invention and unprecedented engineering ambition – bridges, railways, shipbuilding, construction and architecture. The heroes of the age laid much of the foundation for modern scientific and industrial life – Michael Faraday, Charles Wheatstone, Humphry Davy, John Dalton, Isambard Kingdom Brunel, Joseph Whitworth and Charles Darwin. The nineteenth century was not only an age of reason. It was also an age of quantification in which science and engineering set about reducing the world to number. With the rise of science and the burgeoning Industrial Revolution, the need for accurate and convenient numerical calculations mushroomed.

There was one need for tables that was paramount – navigation. Navigators found their position on the open seas from the moon and the stars, and astronomical tables of the kind being checked by the two friends in 1821 were crucial for this purpose. Britain was a leading maritime nation, and accurate navigation was critical to the safety of the fleet both for the protection of the realm and for trade. The stakes were high. Capital, private fortunes and lives were at stake.

The problem was that tables were riddled with errors. Finance, trade, science and navigation were at risk from hidden dangers, and the insecurity of flawed tables undermined the certainties promised and sought by the burgeoning new sciences. When errors were found in published tables a correction sheet of errata was issued and included in the next edition, and these correction sheets give some clues to just how serious the problem was. Dionysius Lardner, professor of natural philosophy and astronomy at London University, and a prolific populariser of science, sought to expose the sorry state of affairs. He inspected a private collection of 140 volumes of tables (probably Babbage's) which had a printed area covering more than sixteen thousand square feet – the size of about six tennis courts. In a random selection of 40 volumes he found no fewer than 3,000 errors acknowledged in the errata sheets. Some of the correction sheets themselves contained errors. Lardner ridiculed the need for errata of errata, and in the case of the *Nautical Almanac* – a standard volume of astronomical tables – he trumpeted the absurdity of errata of errata of errata. Lardner's case was that tables were generically flawed. Yet others argued that they were accurate enough. There was no absolutely certain way of verifying tables, and experts disagreed about whether there was a crisis at all, notwithstanding Lardner's painful exposé.

The problem was not the errors already found and flagged in the correction sheets, but the insecurity of not knowing how many errors remained undetected. Herschel, writing in 1842 to the Chancellor of the Exchequer, Henry Goulburn, captured this deep anxiety of the unknown: 'an undetected error in a logarithmic table is like a sunken rock at sea yet undiscovered, upon which it is impossible to say what wrecks may have taken place'. News of shipwrecks was a constant reminder of the dire consequences to navigation of undetected errors. Scientific magazines buzzed with novel life-saving apparatus about which outlandish claims were made. One device offered a rubber suit with a buoyancy belt and weights to keep the hapless victim vertical. A floating store provided drinking water, food, reading material ('so that he may read the news to pass the time'), cigars, and a pipe and tobacco, as well as torches and rockets to signal the victim's position to rescuers. The kit included metal frames clad with rubber which fitted round the survivor's hands so that, when tired of reading, smoking, eating and drinking while bobbing around, he or she could go places by paddling. Another device was a buoy with sailcloth trousers, gumboots and metal hoops to protect against rocks and voracious fish. The upper section of the buoy was open and large enough to allow some freedom of movement of arms and head, as well as storage for a month's supply of food and water. A concertina cowling like that found on a modern convertible car could be pulled over as a roof in high sea. Yet other devices used pedal power for propulsion and lights for attracting attention.

The preoccupation with shipwrecks was not confined to scientific magazines and reports of sensationally impractical inventions. The *Illustrated London News* carried stark accounts of distressed ships often depicted in dramatic engravings. The foundering of the iron steamer *Brigand* was reported with an artist's rendering and a vivid eyewitness report by a survivor of 'this lamentable catastrophe'. The *Brigand* was only two years old, and had been built at a cost of £32,000 to ply between Liverpool and Bristol. Such reports rarely identified navigational error as the cause of disaster, nor did they specifically incriminate the accuracy of tables. But they were constant public reminders of the dangers of sea voyages, and provided a strong base from which to argue the dread consequences of erratic navigation.

The source of errors in tables was clear: human fallibility. The preparation of mathematical tables was a laborious, tedious and exacting task. For a start, mathematicians devised the formulae and the range of values the tables were to cover. This was the clever part. They then calculated the 'pivotal values' – series of numbers at relatively large intervals. Filling the gaps involved calculating all the values in between (called sub-tabulation). This was a task of awesome drudgery. Each entry in the table had to be calculated by hand, and the production of a complete table demanded the seemingly endless repetition of the same series of arithmetical steps performed on slightly different numbers. To spare themselves, the mathematicians farmed out the routine work to the 'computers' – a common reference to people who performed calculations. (Computers are not the only devices to have human antecedents. Later in the nineteenth century, 'typewriter' referred to a person who typed rather than to the machine itself. It was not unheard of for the boss to elope with his typewriter.)

It was standard practice to give the same task to two computers who would perform the same set of calculations without collaboration. The results, independently prepared in this way, would then be checked against each other. This system of error detection was not foolproof, and it was not unknown for both computers to commit the same mistake. But if independently calculated results were found to be in agreement, this established a high degree of confidence in the correctness of the entry. Any discrepancy in the two results signalled an error and the offending pair of numbers were checked for correctness. This was the stage that Babbage and Herschel had reached when Babbage was driven to proclaim what is perhaps one of the most resonant utterances of the nineteenth century in appealing to 'steam' to execute calculations.

Once checked, results were copied by hand into lists and the manuscripts given to printers for typesetting. Inevitably the process of transcription was itself vulnerable to error. The typesetter then set the results using loose metal type, in preparation for printing. A compositor read each numerical result and selected a separate piece of type for each digit in the number. Pieces of type were laid alongside one another to make up the digits of the desired number, and sets of type were blocked together in a frame to make up a page for use in a printing press. Typesetting acres of numbers is a monotonous task and one that, as ever, is liable to error. A compositor setting text into type can at least recognise meaningful groups of letters making up words. But numbers have no immediate meaning, and there is no intuitive sense of whether or not any one digit has a sensible relationship to the one before or after.

Even when printed sheets of tables finally rolled off the press, the job was not yet done. The printed copy still needed to be proof-read. Anyone faced with a sea of numbers will be daunted by the burden of verifying them. It is not only monumentally tedious, but the process of checking is itself vulnerable to error however conscientiously the task is carried out.

At each of the four stages in the manual production of tables – calculation, transcription, typesetting and proofing – errors wait in ambush. Human frailty is the enemy of accuracy. What confidence could anyone have in the integrity of the outcome?

But what of Babbage's appeal to steam? Britain was in the throes of industrial revolution. Products poured from the new manufactories, and there seemed no limit to the variety and invention of the industrial arts as new materials and processes of manufacture were exploited. Britain led the way and, as Benjamin Disraeli put it in 1838, was fast becoming 'the workshop of the world'. The essayist and historian Thomas Carlyle called the decades of upheaval 'the Age of Machinery'. Science, engineering and the flourishing new technologies held limitless promise. Machines were the obsession of the times, and the extent to which motive power and mechanism permeated life sometimes touched on the absurd. One scientific magazine proclaimed the benefits of a 'new domestic motor'. The illustration shows a woman sitting in a rocking chair darning a sock. Through a system of pulleys, levers and weights, the oscillating chair rocks a cradle with a cherubic sleeping infant, and operates a milk churn at the same time. The reporter noted:

> By this means, it will be observed, the hands of the fair operator are left free for darning stockings, sewing, or other light work while the entire individual is completely utilised. Fathers of large families of girls, Mormons, and others blessed with a superabundance of the gentler sex, are thus afforded an effective method of diverting the latent feminine energy, usually manifested in the pursuit of novels, beaux, embroidery, opera-boxes, and bonnets, into channels of useful and profitable labour.

On an industrial scale the engines of change were not women, but the steam engines which powered the factories. At the time of Babbage's invocation of steam as the salvation of tables from the curse of human fallibility, steam was both the actual and symbolic agent of change and held the promise of prosperity for all.

The encounter with Herschel in 1821 when the two friends met to verify the calculations of the computers was the genesis episode in the history of automatic computation. The account given here is based on the description Babbage wrote some eighteen years after the meeting.

In a separate record written closer to the events, Babbage is less sure whether it was he or Herschel who made the seminal appeal to steam, and comments that, whoever it was, he was in any case half-joking. Despite the differences between Babbage's accounts, the essential features of the encounter are clear. The problem was the daunting handicap of human fallibility to reliable calculation. The solution was machines.

Babbage's interest in mathematics was evident early on. He entered Trinity College, Cambridge, in April 1810, aged eighteen, already a precociously accomplished mathematician, and as a new undergraduate he looked forward to having his curiosity and mathematical puzzlements illuminated by his tutors. To his disappointment he found his teachers a staid lot, stuck in an unchanging curriculum and uninterested in the new Continental theories which excited him. Disaffected, independent-minded and even rebellious, he pursued a programme of study of his own which favoured the works of French mathematicians. Babbage was a radical: he admired Napoleonic France (with which Britain was still at war), decried the unquestioned acceptance of religious doctrine reflected in the inflexible regulation of university life by the Church, and lamented the stagnant state of mathematics in England. Active and spirited, he became one of the instigators of the Analytical Society, which was dedicated to reform English mathematics.

At Cambridge he enjoyed student life to the full. He formed an enduring friendship with John Herschel, who had entered St John's College in 1809, and relished the company of a wide circle of friends. He played chess, took part in all-night sixpenny whist sessions, and bunked lectures and chapel to go sailing on the river with his chums.

Babbage excelled at mathematics, and as Peterhouse's 'crack man' (he migrated from Trinity to Peterhouse in 1812) he was expected to win honours in the final examinations. The mathematics tripos was the major training ground for those destined for the professions and high office, and high honours in the Senate House examinations carried not only prestige but offered a career head start. One of the features of the byzantine university procedures was to rank the top performers in the examinations as 'wranglers'. To earn the status of wrangler bestowed distinction for life.

Candidates had first to qualify to sit the final examinations and there was a complex formal process called the Acts which pre-sorted candidates into categories according to ability, with questions of differing difficulty set for the various groups. The Acts took the form of a formal public debate in which the candidate defended a thesis of his own choice against designated opponents. A moderator adjudicated the disputations and had the power to prohibit arguments regarded as immoral or heretical, and to fail a candidate on the spot for ignorance or inappropriateness. The moderator would signal that a candidate had failed by proclaiming the Latin word *descendas* (literally, 'you will go down').

The thesis Babbage chose to defend in the ordeal of the Acts was that God was a material agent. The moderator, the Reverend Thomas Jephson, judged the proposition to be blasphemous, with disastrous results. A contemporary account paints the scene.

> Babbage . . . wishing most devoutly to prove that God was material, Jephson's Piety received such a violent shock that '*Descendas*' thundered from his lips . . . All Peterhouse was in an uproar, when the direful news came that their crack man had got a *descendas*.

Herschel was unsympathetic, telling his mortified friend that he deserved everything he got. They argued and Babbage called Herschel a 'blockhead'. 'Herschel, you are as great an ass as myself,' shouted the distraught Babbage, and stormed out.

It is difficult to know whether Babbage's choice of proposition was a principled act of defiance against non-secular education or a lapse of political judgement. He was not alone in his anti-religionist views. Herschel was a kindred spirit, but he and others were more prudent in their public expression. Babbage 'gulphed it' – he was awarded an ordinary degree (in 1814) without examination. His ill-judged proposition perhaps played its part, or he may himself have chosen not to compete for honours, as was the entitlement of those who had already proven their abilities. The more temperate Herschel, on the other hand, graduated in 1813 as senior wrangler and first Smith's Prizeman, a coveted award for proficiency in mathematics established by the will of Robert Smith, master of Trinity, who died in 1768. Herschel went on to become a Fellow of St John's College and like his father he distinguished himself as an astronomer. In time he became a model ambassador for science. Babbage never once referred in his writings to this ignominious episode, and always recalled his student days with a warmth innocent of rancour. His *descendas* was not the last episode of self-destructive public protest. 'The ties which connect me with Cambridge,' he wrote, 'are indeed of no ordinary kind.'

The Babbages were a well-established Devonshire family, and during vacations Charles returned to his childhood stomping ground. It was probably during a visit in summer 1811 to Teignmouth that he met and started courting Georgiana Whitmore, and the couple became engaged in summer 1812. Babbage's father, Benjamin, disapproved. Benjamin, a well-to-do London banker of the firm Praeds, Mackworth & Newcombe, thought it imprudent for his son to marry without means. But Charles was headstrong, and his relationship with his father was in any event not a happy one. On the evening of 1 July 1814 he went off in a chaise to

Teignmouth, taking his former tutor, who had taken holy orders and was therefore able to perform the marriage ceremony, and the happy couple were wed the next morning. Babbage's defiance precipitated an almighty row. A month after the marriage, he wrote to Herschel:

> I am married and have quarrelled with my father. He has no rational reason whatever; he has not one objection to my wife in any respect. But he hates the abstract idea of marriage and is uncommonly fond of money.

Herschel was shocked at his friend's apparently casual attitude to his plight:

> I am married and quarrelled with my father – Good God Babbage – how is it possible for a man to calmly sit down and pen those two sentences – add a few more which look like self-justification – and pass off to functional equations.

Herschel adds a curious postscript: 'You have not informed me who the lady is? I presume it can be no secret.' The two had been in near-daily contact while students in Cambridge. Babbage had been engaged for two years before graduating, yet Herschel had been kept in the dark all this while about his friend's love life.

Babbage replied to Herschel with a bitter attack on his father, and the letter is rare in its level of personal exposure:

> You seem to have a great horror at my having quarrelled with my father . . . but . . . you must know a little of his character before you judge. My father is not

> much more than sixty, very infirm, tottering perhaps on the brink of the grave. The greater part of his property he has acquired himself during years of industry; but with it he has acquired the most rooted habits of suspicion . . . He is stern, inflexible and reserved, perfectly just, sometimes liberal, never generous, is uncultivated except perhaps by an acquaintance with English Literature and History. But whatever may be his good qualities they are more than counterbalanced by an accompaniment for which not wealth nor talents nor the most exalted intellectual faculties can compensate – a temper the most horrible which can be conceived.
>
> Seeking the happiness of no earthly being he lives without a friend. A tyrant in his family his presence occasions silence and gloom . . . Tormenting himself and all connected with him he deserves to be miserable. Can such a man be loved? It is *impossible* . . . I too do what I think is right and should esteem it treason to myself to sacrifice my own happiness in the *vain* attempt of pleasing such a character . . . this is not the hasty sketch of irritated feelings. I am not angry with, I pity such a being.

Even allowing for the anger that may have temporarily distorted Babbage's view of his father, the domestic picture is a bleak one. However, he adored his mother, Elizabeth, whom he credits with tolerating Benjamin out of a sense of religious duty.

With their newly born son, Benjamin Herschel (known always as Herschel), Charles and Georgiana moved to London and in September 1815 took up residence in a small house, at 5 Devonshire Street, off Portland Place. The colossally expensive and protracted war with France had ended in triumph at

Waterloo in June that year, and the mood in England was momentarily upbeat. There was the prospect that income tax, introduced as an emergency measure to fund the wars, would be abolished. Civil works flourished, and seemed to symbolise a new prosperity after the anxieties of the Napoleonic Wars. In London, Regent Street was being built, and new bridges at Waterloo and Southwark would soon span the Thames. The young Babbage family came to London in bustling times of change.

Charles's father had promised him an allowance of £300 a year, a pledge that was honoured despite Benjamin's hostility to the marriage. With an additional £150 per year from rents, the Babbages were moderately well off and able to enjoy a modestly comfortable life on a limited but secure income. Babbage had accepted his father's allowance; he had little choice. But partly through pride and partly through need he set about landing a paid job. The Church was one possibility. As a clergyman he would have a livelihood and the leisure to pursue his mathematical and philosophical interests, but without the comfort of additional funds from his father, the prospect of which was now in doubt, he gave this up. University posts were another option, and he applied for at least two professorships, though without success. His reputation as a published mathematician and a glittering list of referees seemed to count for little in a culture in which appointments were primarily through personal and political patronage, and Babbage wrote of his grievances at being rejected on grounds other than merit. He suspected that his liberal politics may well have prejudiced his chances. Gulphing at Cambridge and graduating without honours could not have helped. Herschel, in the meantime, was reading law, as were others of Babbage's friends.

He also had his share of bad luck. He was offered the

position of director and actuary of a start-up life assurance company with the prospect of an annual income of £2,500, a thumping improvement on his father's grudging allowance. He absorbed himself with vigour in actuarial statistics and computed a new set of tables relating premiums to life expectancy. But the venture was abandoned the day before the scheduled launch, possibly following an argument between Babbage and the other directors. The couple's financial situation was further eased on the death of Georgiana's father in 1816. Charles's bank book shows that a budget of £126 in 1814 grew to a sizeable £2,355 by 1822. Still he sought a paying post.

Science was not an established profession and offered no secure career path. There were few civil scientific appointments, and for the most part scientific activity was conducted by gentlemen of independent means with the leisure to pursue their amateur interests. Although science flourished in the early decades of the nineteenth century, in cultural terms it was still a Cinderella activity and had yet to seriously challenge the accepted view of the world offered by religious teaching. In the 1820s science enjoyed nothing like the authority it now automatically commands. It was not yet recognised as a distinct mode of enquiry, and the now widely accepted notion that it offered some privileged access to certainty had yet to be earned. The word 'scientist' was not coined until 1833, when a Cambridge don, the formidable William Whewell, used the term to distinguish those concerned to explore the material world from those engaged in literary, religious, moral and philosophical pursuits. Scientists were only just becoming identified collectively as a distinctive occupational tribe.

The hub of scientific activity was the Royal Society, founded by royal charter in 1662. The Society was as much a social club for country gents desirous of a London base as it was the headquarters of scientific life. Entrance requirements were undemanding, though the Society's meetings and published *Transactions* were the talking points of the scientific community, and the Society's councils and committees were the political power base of scientific life in England.

While he was seeking paid employment, Babbage pressed on with his scientific interests, and once in London he quickly established himself in scientific circles. In 1815 he gave a series of twelve general lectures on astronomy at the Royal Institution in Albemarle Street, and was elected a Fellow of the Royal Society the following year, with John and William Herschel as two of his nominees. He was one of the legendary group that dined at the Freemason's Tavern in Lincoln's Inn Fields on 12 January 1820, when the founding of the Astronomical Society was finally resolved. Babbage would serve as one of the first of the Society's secretaries.

His reputation was enhanced by a string of mathematical papers, and his intellectual vigour, his publications and his friends placed him in the thick of scientific life. By the time of his meeting with Herschel in 1821 we find Babbage happily married and enjoying the life of a gentleman philosopher in Regency London. His circle of friends and scientific associates frequently called on one another, dined at one another's houses, shared their discoveries and engaged in the politics, scientific intrigue and gossip of the day.

Soon after the genesis meeting with Herschel in 1821, the friends went off together on a trip to the Italian Alps. They would have had ample opportunity to discuss calculation and table-making while climbing. Whether or not they did is unclear. But the episode of tabular errors haunted Babbage, and he was seized with the challenge and fascination of calculation by machinery. His absorption in the problem and the sustained mental application to its difficulties was such that once back in England he became ill and was advised to take a rest. He went to stay with Herschel in Slough. As Babbage himself recounts:

> The excitement of the enquiry had an unfavourable effect upon my bodily health; and I was recommended to abstain entirely for a time from all thought of the 'Calculating Engine'. I accordingly went to visit my friend at Slough and continued to obey strictly the injunctions of my medical adviser.

But he was hooked. When Herschel went up to London to attend a meeting of the Board of Longitude, Babbage took advantage of his friend's absence to explore how a machine might perform arithmetic calculations. By the time Herschel returned that evening, Babbage had a preliminary scheme to discuss. Herschel had seen nothing like it.

Babbage was exhilarated by this early progress and had little idea of what lay ahead. In later years he recalled his early innocence:

> It was certainly fortunate for me, both at this period, and at many other times, that I had no sufficiently distinct view of the multitude of difficulties, both practical and moral, which were destined to attend its course. Had these not opened upon me by degrees, I might perhaps never have ventured on its execution.

On his return to London he continued to explore the design and set about constructing a small model. To protect the invention he farmed out the manufacture of parts to different workmen and assembled the device after they had left – an uncharacteristic act of secrecy. By the spring of 1822 Babbage had a small working model of his first design. He made no attempt to build any part of the printer, though he explored several designs for setting type automatically.

We have seen that the risk of errors plagued all four stages in the preparation of printed tables: calculation, transcription, typesetting and proof-reading. Babbage's grand scheme was that his calculating engine would eliminate all four sources of error at a stroke. Infallible machines would solve all. The 'unerring certainty of mechanical means', as Lardner put it, would guarantee absolute accuracy of calculation. Babbage reasoned that errors of typesetting would be eliminated if the machine could automatically impress the results of the calculation into soft metal or into papier mâché. These 'stereotypes', as they were called, could then be used as moulds for making printing plates from which tables could be printed in volume using a conventional printing press without an error-prone human intermediary. By mechanically coupling the printing and stereotyping apparatus to the calculating mechanism, the supposed infallibility of machinery would not only guarantee the integrity of the calculation itself but would also extend to the printed record. Apart from the role of supplying initial values to start the calculation, the fallible human would be eliminated from the process.

For all their ingeniousness, the ornate calculating devices that decorated the drawing rooms of the elite were unreliable. This was not their only drawback. To put them to any useful purpose the operator needed to enter numbers on dials, operate the machine, and then write down the results – operations susceptible to error, as we have seen. These machines were not automatic but relied on the continuous informed intervention of the operator. The new calculator – the arithmometer – introduced by Thomas de Colmar in 1820 was also dogged by unreliability. Robust and successful as it was later to become, the arithmometer also relied on the manual manipulation by the operator of dials, sliders and a movable carriage like that on a typewriter, and, as with the earlier devices, results needed to be copied out by hand with the attendant risk of errors.

There was another restriction. The devices of the time were limited in the size of the numbers they could handle. The problem here was how to carry from units to tens, from tens to hundreds, and so on when a digit exceeded '10'. The worst case arises when the number is 999,999 and a '1' is added. The '1' needs to be carried to each digit position, rippling through the wheels in turn in a domino effect. The difficulty with the calculators of the day which used dials to enter numbers was that the force required to advance all the digit wheels in a 'domino carry' is derived from the one movement which adds the '1' to the last position. With calculators made of wood, ivory and soft workable metals, the strength of the materials placed an upper limit on how many digits could be driven. The devices of the time could typically handle numbers with six or sometimes eight digits, but no more.

Babbage's scheme set out to remove the handicap of manual operation as well as solve the problem of limited accuracy. Machines, like humans, find it easier to add than to multiply or divide. Designing a mechanism to perform automatic multiplication and division is a formidable task – one which Babbage did not accomplish until much later. Instead he used a simplifying principle called the 'method of finite differences', a technique which avoids the need for multiplication and division in calculating the values of a certain class of mathematical expressions. The method was well known at the time, and was routinely used by human computers preparing mathematical tables. The beauty of the method of differences is that it uses only arithmetical addition, and removes the need for multiplication and division in calculating the values of the general class of mathematical expressions called polynomials. All that is required in using the method of differences to generate successive values in the table is a series of repeated additions which can be performed by simple gear wheels, and the difficult problem of getting a machine to multiply and divide is avoided.

Building an engine that can calculate only one class of mathematical expression – in this case polynomials – sounds like a crippling restriction. But polynomials have a crucial property which generalises their use. Almost any regular 'well-behaved' mathematical function, logarithmic and trigonometric functions, for example, can be approximated by a polynomial to any required accuracy within a fixed interval. So while the machine was confined to tabulating this one function, being able to do so extended its use to a very wide range of other mathematical applications.

The method of differences also solved the problem of proof-reading. The method uses the last result to calculate the next one – each new value in the table depends on the one immediately preceding. The very last value therefore depends on all its predecessors. *Voilà*! Instead of verifying each and every digit of each number, all the proof-reader needs to do is check the correctness of the last result to verify the whole table. If the last result is correct, then there is a high degree of confidence in the reliability of all the preceding results, on which it depends. So the method of differences, a powerful general technique, brought the problem of tabular calculation within the compass of machinery by reducing computational demand to simple repeated addition, and with the added advantage of solving the difficulty of verification.

In Babbage's calculating machine, numbers are represented by gear wheels inscribed with the figures '0' to '9'. Each digit of a number had its own wheel, and the value of a digit was represented by the amount by which the wheel rotated. The machine automatically moved the wheels to perform repeated addition according to the method of differences. The operator set the starting values on the wheels by hand, and each cycle of the engine then produced the next result in the table.

Babbage was not the first to suggest a printing calculator, nor was he the first to propose the method of differences as a suitable principle on which to base a mechanised calculation. This distinction goes to one Johann Helfrich Müller, a German engineer and master builder who described his idea in a letter to a colleague in 1784:

> I would in the future make a machine, which would simultaneously print in printer's ink on paper any arbitrary arithmetical progression . . . and which would halt of its own accord, when the side of the paper was full up. After setting the first figure, all one has to do is to turn a handle.

The machine was never built. A booklet published two years later specifically mentions Müller's idea of using the differencing principle to produce 'a whole series of numbers' – this thirty-five years before the genesis of Babbage's scheme. But there was no attempt at a design or any outline of how this might be achieved.

The question of whether Babbage had any knowledge of Müller's work is a tantalising one. Herschel translated into English key selections from Müller's published account for Babbage, but the manuscript is undated. It is impossible to know whether Babbage had read Müller's account before his own proposal. But it is fair to say that, had he done so, concealment would have been uncharacteristic. He was a stickler for propriety and a fierce defender of moral probity. He campaigned, for example, to secure recognition for the prediction by John Couch Adams of the existence of the planet Neptune, which he thought improperly acknowledged, and he rarely lost an opportunity to express righteous indignation about issues that offended his sense of fairness. By his conduct and public declarations it would have been wretched of him not to give Müller credit if he thought it due. The balance of probability favours his ignorance of Müller's work, and suggests that each originated their ideas independently.

Babbage had taken a huge step. His world was full of promise. But by his own admission he had little inkling of where his quest would lead.

Chapter 2

A PERSONAL QUESTION

I think it likely that he lives in a sort of dream as to its utility.

George Biddell Airy, Astronomer Royal, 1842

The small working model completed in the spring of 1822 was a novelty in Babbage's circle. The leading lights of science, many of whom were Babbage's personal friends, came to see the new machine and to hear Babbage expound upon its wonders. The time had come to announce the invention, and he sent a brief notice to the Astronomical Society in June 1822 in which he described his machine as producing results 'almost as rapidly as an assistant can write them down'.

By way of wider official announcement, Babbage wrote an open letter to Sir Humphry Davy, President of the Royal Society. The letter refers to 'the most stupendous monument of arithmetical calculation which the world has yet produced'. This is not a reference to Babbage's own invention, but a tribute to the largest and most ambitious table-making project ever undertaken by manual means: the late-eighteenth-century production of 'cadastral' tables in France for accurate land survey so that property could be appropriately taxed. The man charged with this project was an engineer, Baron Gaspard Clair François Marie Riche de Prony, who set about producing a definitive set of logarithmic and trigonometric tables using the metric system newly adopted in Napoleonic France. The full set of tables runs to eighteen volumes. Babbage had almost certainly seen the tables during an earlier visit to Paris, and estimated that one table of logarithms alone contained eight million figures.

It was not only the scale of the venture that captured Babbage's attention but the organisation of the work. De Prony followed the principle of the division of labour developed by the political economist Adam Smith in his famous *A Treatise on the Wealth of Nations* published in 1776. De Prony split the human computers into three teams, each of which was to undertake a different level of work. The first team consisted of five or six top-ranking mathematicians, including Adrien Marie Legendre and Lazare Carnot, who chose the formulae to be used. The second team of about eight human calculators worked out the main 'pivotal' values. These are like fence posts set at large intervals with gaps between them waiting for slats to fill the spaces in between. This second group also provided the tables of differences from which the intermediate values could be calculated, and later checked the two sets of results independently calculated by the third team.

The last team was the largest and consisted of sixty to eighty computers who had the laborious task of carrying out the repeated additions and subtractions to produce the missing intermediate values. The computers who did the lowest level of work were mostly capable of only elementary arithmetic, and required 'the least knowledge and by far the greatest exertions'. Babbage observed that when the maths experts higher up in the hierarchy of skills tried this routine of sub-tabulation, they made more mistakes than the drudges below them.

Many of these low-level computers were out-of-work hair-dressers. With the guillotining of the aristocracy, the hairdressing trade, which had tended the coiffures of the elite, was in recession. The elaborate hairstyles had become a loathed symbol of the defunct pre-revolutionary regime, and many hairdressers turned their hand to rudimentary arith-metic.

With the benefit of a calculating engine, argued Babbage in his letter to Sir Humphry, de Prony could have done away with the entire third group of computers, and further reduced the staffing levels of the other two groups. It seems that every-one had it in for hairdressers. Babbage estimated that if a machine were used the task would require only twelve calcu-lators instead of over ninety. He did not consider the capital cost of an engine which was massively high compared to the low rates for unemployed hairdressers; these painful debates were still to come.

The model itself received a mixed reception. Many of those who saw it were impressed, others less so. Babbage recorded his visitors' reaction in his journal: 'May 10 [1822]. My calcu-lating engine is nearly finished. Of those whom I have made acquainted with the principles many think too little of it. Of those ignorant of them the greater part think far too much.' One heavily underlined entry in his journal records the reac-tion of his friend and fellow scientist, William Hyde Wollaston, who had earlier encouraged him in his engine pur-suits:

> He appointed to visit me at my house in an hour to see the engine. Dr. W. examined it and I explained all its

> parts and then worked it and after about an hour and a half the result of his opinion was expressed in these words. '*All this is very pretty but I do not see how it can be rendered productive.*'

It is hard to know whether Babbage was stung by this remark or whether he took it in the spirit of friendly joshing. Whatever Wollaston meant by the comment, it was a signal to Babbage that the ingenuity of the invention did not automatically protect it from questions about its practical usefulness.

The letter from Babbage to Sir Humphry ends in a curious way:

> Whether I shall construct a larger engine . . . will in great measure depend on the nature of the encouragement I may receive . . . success is no longer doubtful. It must however be attained at a very considerable expense, which would not probably be replaced, by the works it might produce . . . and which is an undertaking I should feel unwilling to commence, as altogether foreign to my habits and pursuits.

He does not ask outright for financial support, but there is the hint that he is not immune to persuasion.

Babbage sent privately printed copies of the letter to colleagues and influential friends. The political machinery lurched into motion, and the engine issue began to acquire a momentum all its own. Davies Gilbert, a colleague of Babbage's who was Vice President of the Royal Society and, perhaps more importantly, a Member of Parliament supportive of scientific interests in the House of Commons, lobbied Sir Robert Peel who was then First Lord of the Treasury. Gilbert's idea was that a full-sized version of Babbage's engine should be constructed at public expense, and he suggested that the Royal Society should be consulted for a professional opinion. The Royal Society was a private and self-governing body, formally independent of Parliament, and its members and committees acted from time to time as scientific advisers to government with apparent neutrality. Sir Robert wanted advice before he risked raising the matter in the House of Commons, and in March 1823 he consulted his friend and close adviser John Croker, Secretary to the Admiralty.

Peel was sceptical, and his letter to Croker suggests with some facetiousness that the notion of a 'scientific automaton' was far-fetched. At the Admiralty, Croker would have been sensitive to the potential benefits of the machine for table-making for astronomical navigation, and was already aware that the Board of Longitude had considered a one-off award of £500 to Babbage for the invention to 'reward his ingenuity, encourage his zeal, and repay his expenses'. Croker acknowledged the potential of the machine but was cautious about financial support from the public purse:

> I cannot but admit the possibility, nay the probability, that important consequences may be ultimately derived from Mr. Babbage's principle . . . As to Mr. Gilbert's proposition of having a new machine constructed, I am rather inclined . . . to doubt whether that would be the most useful application of public money.

Croker did, however, endorse Gilbert's suggestion to refer to the Royal Society, and the Treasury requested that it 'be favoured with the opinion of the Royal Society on the merits and utility' of Babbage's invention.

A committee was duly convened to advise on the prospects for Babbage's engine. The findings of the committee, which reported on 1 May 1823, were favourable. It commended the 'great talents and ingenuity' of the engine's inventor and concluded that Babbage 'was highly deserving of public encouragement in the prosecution of his arduous undertaking'. But there were divisions in the Royal Society about the utility of the machine. In a confidential letter written two decades later by the Astronomer Royal, George Biddell Airy, to the Chancellor of the Exchequer, Henry Goulburn, Airy gives some insight into the situation behind the scenes. First he comments on the membership of the Committee: 'These persons were all private friends and admirers of Mr. Babbage: and . . . I cannot help thinking that they were a little blinded by the ingenuity of their friend's invention.'

Airy's allegation is not without foundation. Of the twelve members of the 1823 committee, at least seven were Babbage supporters of one sort or another – close friends, men of science involved with Babbage in collaborative projects, or associates who shared his political sympathies. His allies included John Herschel, Davies Gilbert, William Hyde Wollaston, Francis Baily, Marc Isambard Brunel, Thomas Frederick Colby, and Henry Kater. Despite the presence of Babbage's allies, support for the project was not unanimous: 'Their Report was very favourable. When the Report was discussed by the Council of the Royal Society, it was boldly stated by Dr. Young . . . that, if finished, [the machine] would be useless.'

Thomas Young was no novice when it came to tables. He served for a long period as superintendent of the *Nautical Almanac*, and was secretary of the reconstituted Board of Longitude from 1811 to 1829. He made it clear that he had no doubt that the engine could be made, but maintained that it would be a better bet to invest the probable cost of the engine and use the dividends to pay human calculators to do the computation. Airy recalled Babbage's reaction:

> For this [the statement that the engine would be useless], he [Young] was regarded by Mr. Babbage with the most intense hatred . . . Mr. Babbage made the approval of the machine a personal question. In consequence of this, I, and I believe other persons, have carefully abstained for several years from alluding to it in his presence. I think it likely that he lives in a sort of dream as to its utility.

Airy portrayed Babbage as irrationally defensive of his engine and hypersensitive to any view that compromised its desirability or use. But Airy's revelations were written much later. At the time the committee's official report was a ringing endorsement for Babbage's engine despite the high passions and disagreements behind closed doors. Fortune smiled and on 13 July 1823 the Secretary of the Astronomical Society notified Babbage that he was to be presented with the Society's first Gold Medal, the first public recognition for his invention. It seems that Herschel and Edward Bromhead, a cohort from Babbage's student days, had done some lobbying on their friend's behalf and helped fortune along.

At the same time funding fell into his lap. In a private interview in June of that year with the Chancellor of the Exchequer, John Robinson, Babbage was informed that the Government would provide financial support to construct an engine, and a Treasury warrant for £1,500 was issued as the first payment. Babbage's commission was to 'bring to perfection a machine invented by him for the construction of numerical tables'. Babbage was overjoyed and wrote to Herschel with the news saying that he was optimistic that in a few years his completed engine would produce 'logarithmic tables as cheap as potatoes'. (The reference to potatoes is unfortunate. When the potato blight struck central and Western Europe in 1845, the engine was still unfinished and had cost the Treasury a goodly fortune.)

The way seemed clear for Babbage to undertake the design of a larger machine. Flush with funds, he set about designing a properly engineered working Difference Engine large enough to perform calculations for the production of tables. His first experimental model had been poorly made. So that the quality of the engineering did not compromise the more ambitious machine, he studied mechanical workshops and the manufacturing arts to establish what was achievable using existing practice. He travelled the country, visiting the major industrial centres in the north of England and Scotland, touring manufactories, craft and engineering workshops to learn what he could of the mechanical arts, machinery, materials and the state of development of the machine tool industry.

As a well-to-do gentleman he could have stayed in grand hotels or with titled acquaintances. Instead he used local inns, partly out of unnecessary thrift, but more particularly because it brought him into contact with tradesmen whom he could tap for know-how about techniques, developments and tooling. On one trip he spent a week incognito at a hotel used by tradesmen in Sheffield. A travel acquaintance told him at breakfast one morning that after he had gone to bed the night before, the company had fallen to guessing what trade he was in. Babbage recalled:

> 'The tall gentleman in the corner,' said my informant, 'maintained that you were in the hardware line; whilst

> the fat gentleman who sat next to you at supper was quite sure that you were in the spirit trade. Another of the party declared that they were both mistaken . . . and that you were travelling for a great iron-master.' 'Well,' said I, 'you, I presume, knew my vocation better than our friends.' – 'Yes,' said my informant, 'I knew perfectly well that you were in the Nottingham lace trade.'

Babbage was hugely flattered to be mistaken by travelling tradesmen for one of their own, but the next day his cover was blown by the arrival of Lord Fitzwilliam's liveried groom with an invitation to spend a week at the family estate (which he duly did).

The industrial tours were arduous and often uncomfortable. In the 1820s rail travel was virtually unknown, and a journey of sixty or eighty miles by carriage was not to be undertaken lightly. At least one of Babbage's trips was lightened by the company of his wife, Georgiana, who joined him on his tour through England and Scotland in 1823.

His extensive survey of the industrial arts of Britain led Babbage to the tough conclusion that making the engine would stretch existing technology to its limits. To produce certain intricate parts he would have to devise and construct many of the tools himself. A less vigorous and original thinker might have been deterred. But Babbage converted the stables at the back of his house in Devonshire Street into workshops and staffed them with hired workmen under his personal direction. It was soon evident, however, that the intricacy of the engine was beyond his makeshift facilities. He needed specialist tools and the knowledge and skills of a professional machinist and engineer.

While Babbage was casting around for specialist help, Marc

Isambard Brunel, a renowned civil engineer and Babbage's friend, recommended Joseph Clement, a highly skilled toolmaker as well as a first-rate draughtsman. It was rare to find the skills of a machinist and draughtsman in one person and rarer still to find someone like Clement, who excelled at both and was well able to produce detailed drawings of elaborate parts and meticulous illustrations of machines. Time and again this unusual combination secured him advantage. In 1813 at the age of thirty-four and with £100 of savings, he had moved from Glasgow to London to advance himself. He was dropped by coach in Holborn and immediately offered his services to Alexander Galloway, proprietor of a local machine shop. 'What can you do?' asked Galloway. 'I can work at the forge.' 'Anything else?' 'I can turn,' replied Clement. 'What else?' 'I can draw.' 'What!' said Galloway. 'Can you draw? Then I will engage you.' Galloway, it seems, was a pretty ropy engineer. A roof he designed collapsed and killed eight or ten of his men. Clement found the quality of the workshop's tools poor and, to the admiration of his co-workers, he forged his own. His skill, dexterity and ingenuity quickly marked him out, and he rapidly rose above the rest.

Under Galloway, Clement was earning a guinea a week. Aware of the value of money and the quality of his own work, he approached Joseph Bramah, one of the great pioneering engineers of the day, taking with him examples of his drawing. He was hired on the spot and soon commanded three guineas a week as chief draughtsman and superintendent of Bramah's works at Pimlico. He moved shortly after to become chief draughtsman for Henry Maudsley, another of England's outstanding engineers. His dogged trust in his own abilities had paid off. His draughtsmanship was unrivalled and he was equal to the best at Bramah's and Maudsley's when it came to precision machining and toolmaking.

With £500 of savings he set up on his own in 1817, taking premises at Prospect Place, Lambeth, just south of the river, and soon established a reputation for precision work and fine finishing. He was forever making new machines, improving existing ones and innovating techniques of manufacture. He also had an eye for a good earner. He devised a planing machine which was able to take larger workpieces than any other of its kind. Such was the demand that it was often kept running night and day. At eighteen shillings a square foot of planed work, the planing machine brought in £10 in a twelve-hour day, and for a while this was Clement's principal source of income. Dedicated to his tools, he lived above his workshop and was rarely tempted away from his premises.

Clement was blunt and gruff to the point of belligerence. There are no surviving portraits of him, but he is described as a 'heavy-browed man, without any polish or manner of speech'. As the son of a poor handloom weaver he was put to work when young and had only a rudimentary education. His written bills tended to be abrupt, near-illiterate demands for money. But he knew his worth, and was quite unintimidated by rich or illustrious customers.

On one occasion Isambard Kingdom Brunel, engineer to the Great Western Railway, asked if Clement could supply a louder steam whistle for his locomotives to replace the ones in use, which were reedy and feeble. Clement examined the sample brought by Brunel and scorned it as 'mere tallow chandler's work'. He devised the tools he needed for a new whistle and when this was tried it screamed with a joyously deafening hoot heard miles away. Brunel ordered a hundred of them, and was well pleased – until he got the bill. 'This is six times what we now pay,' he protested. 'That may be,' replied Clement, 'but mine are more than six times better. You ordered a first-rate article, and you must be content to pay for it.' The dispute was referred to an arbitrator, who ruled in Clement's favour.

This was not the only instance of large bills for work of unparalleled precision. An American ordered a machine screw of the kind used on lathes to be made 'in the best possible manner'. Clement was happy to oblige and cut the screw with mathematical accuracy. The bill came to hundreds of pounds and staggered the customer, who had expected to pay no more than £20. Again the arbitration process found in Clement's favour and the bill had to be paid.

If it was precision engineering you needed, new machine tools and exquisite engineering draughtsmanship, then Clement was your man – provided, that is, you had the money. This was precisely the Babbage cocktail in the mid-1820s. He had a mandate from government to build his engine, money in the bank, and a colossal task that would make unprecedented demands on the state of the mechanical arts in design, precision and intricacy. On Brunel's recommendation Babbage hired Clement, and a partnership of mixed fortunes began. The men were worlds apart in class, education and means. Their lives intersected in the mechanisms of a calculating engine.

With Clement hired, Babbage began in earnest the design of the first fully engineered difference engine, which he called Difference Engine No. 1. He designed and sketched the mechanisms, and Clement drafted them. Babbage would invariably consult Clement on how he should go about making the parts. Babbage was the inspirational source of the concept and function of the elaborate mechanisms, Clement the source of practical skill and know-how. The collaboration was close, and it is impossible to establish how much of the technical detail of the engine was Clement's and how much Babbage's.

Precision was an incessant anxiety. Babbage demanded the highest precision possible, and his demands often went beyond achievable limits of skill and machinery. Where existing practice was inadequate, Babbage and Clement devised new tools and modified machines that were to hand. In meeting the needs of the engine they made significant contributions to the development of machine tools and manufacturing practice. But precision came at a cost, and the bills rolled in.

There was a profounder and more subtle challenge: the need to produce large numbers of similar parts. Babbage's calculating engines were like no other machines of the time. Their massive cast-iron frames identify them as part of the great age of machinery, but they do not resemble clocks, steam engines or textile machines, the devices we associate most closely with the early half of the nineteenth century. What is distinctive about Babbage's designs is that they call for more repeated parts than any device designed up to that time – hundreds and sometimes thousands of near-identical parts.

While the manually operated desktop calculators of the time boasted six or perhaps eight digits, Babbage's engines were designed to handle numbers with twenty, thirty and even fifty digits. One of his later designs for a fifty-digit machine has facilities for double-precision results which allow one hundred significant digits in the answers. This was a staggeringly ambitious numerical range, and way beyond anything previously proposed. Even a modest-sized Babbage engine has several hundred similar number wheels, and one of the most dramatic visual features of all Babbage's engines is the overall impression of a regular array of indistinguishable parts.

The reason for such a high level of repetition of near-identical components lies in the way the engines store and manipulate numbers. In our familiar decimal system the ordinary counting numbers are represented by the ten symbols '0' to '9'. A largish number, say '9,542', has four separate digits. The '2' represents the number of units, the '4' the number of tens, and so on. In Babbage's engines each digit is stored on a gear wheel that he called a 'figure wheel', inscribed with the numbers '0' to '9', and each individual digit in a number has its own separate identical figure wheel.

So the number '9,542', which has four digits, will be held in the machine by four separate figure wheels in a vertical column with the units at the bottom, tens next up, and so on. In this example the units wheel at the bottom of the stack will be turned two teeth to represent '2'; the tens figure, which is the next one up, will be turned four teeth to represent '4'; and so on. In this way the value of a number is represented by a vertical stack of inscribed figure wheels, each one rotated according to the value of the digit it represents. The early designs for Difference Engine No. 1 had six columns, each with at least twelve figure wheels. By 1830 this had grown to seven columns of figure wheels and sixteen digits. By the 1840s his Difference Engine No. 2 boasted eight columns each with thirty-one figure wheels. So even the early designs called for near-identical parts in their tens and hundreds, rather than the one-off or small quantities typical of production at the time.

Babbage had the misfortune to conceive of his engine when manufacturing technology was in transition between craft and mass-production traditions, before there were established methods for the production of similar parts in quantity. Components were made one at a time, and each was slightly different. By careful comparison of dimensions parts could be made nearly identical by finishing with hand tools or by incremental machining. They could be 'made to fit' by careful filing, tweaking and reaming by hand. But the process was labour intensive, time consuming and therefore costly. Babbage was handicapped by the lack of manufacturing processes that feature *inherent* repeatability, such as stamping sheet metal with shaped punches and dies; precision pressure diecasting, for which molten metal is injected into shaped moulds; or copying by automatic machines.

Even given these limitations it would seem possible to accelerate the production of parts by having several hands make the copies of the same part at the same time, or farming manufacture out to several workshops. The ability to make multiple copies at the same time by 'paralleling up' the production process is something we now take for granted. But in the 1820s the problem was that there was practically no standardisation in manufacturing, and this kept things slow and thwarted progress. Each mechanical engineer had his own taps and dies used for cutting screws, and no two were the same. Each lathe had a different master screw used as the pattern for cutting screw threads, so a screw cut on one lathe would be different from one cut on its neighbour, even in the same workshop. Without standardisation, parts produced on different machines were simply not interchangeable. You could not expect a nut cut in Manchester to fit a bolt made in London, except by happy accident. So however much Babbage might have wished to speed up manufacture, he was bound by the practice of the day to use only one component supplier – in this case Joseph Clement – and had to live with the bottlenecks and frustrations of grindingly slow progress.

The demand for repeat parts for Babbage's engine had a far-reaching effect on production engineering in later years. If a machine made elsewhere needed a new bolt, the thread cut into the machine, even if undamaged, had to be drilled out and retapped to match the new component. Clement saw the waste and trouble of this, and in about 1828 settled on a fixed number of threads in a given length of screw. The scheme was not an immediate success. But a journeyman working for Clement on Babbage's engine, Joseph Whitworth, played a major role in the adoption of standards two decades later through the introduction of the eponymous standard Whitworth screw thread which dominated engineering practice for over a century. Impressive as Clement's contributions are to the development of machine tools, his contribution to standardisation probably had more impact on mechanical engineering than did his ingenious machines.

By 1826 Babbage was wholly absorbed in the design and development of his Difference Engine. In November of that year he wrote to the Master of Trinity College and Vice Chancellor of Cambridge University with the first hints of personal sacrifice to the project:

> I did not pledge myself to devote my whole time exclusively to this project, yet I feel that the liberal and very handsome manner in which I was received at the Treasury would be but ill returned if I were to allow any other agreements to impede its progress. I have hitherto given up everything for this object, situations far more lucrative . . . have been sacrificed, and I should not wish to change these sentiments now that it is approaching, I hope, to a successful termination.

Babbage's assessment that he was close to completion was wildly optimistic, and the theme of unrewarded personal sacrifice was to become a running motif. But meanwhile he and Clement laboured on in an earnest and glorious collaboration in which Babbage shared his time between his own workshop in Devonshire Street and Clement's works, four miles away on the other side of the Thames. The collaboration between the two men did the advancement of engineering no end of good. The conception, scope and detail of what they undertook is extraordinary. The Engine was massively larger than anything that had gone before – eight feet high, seven feet long and three feet deep, weighing an estimated fifteen tons. The design called for some 25,000 separate parts, equally split between the calculating section and the printer. Samuel Smiles, an industrial biographer of the time, records that the drawings for a portion of the calculating section alone covered four hundred square feet in area. By any standards the undertaking was a giant leap in logical conception, physical scale, and complexity.

Chapter 3

TRAGEDY AND DECLINE

Who the deuce ever did anything worth naming without sacrifice?

John Herschel, 1830

Babbage was absorbed and enthralled by his Engine, though not to the exclusion of his many other interests. His advice was sought on novel and obscure problems. In one instance he was asked to evaluate a system for winning at roulette. A syndicate had devised the betting scheme for use in Continental gambling dens, and was looking for £5,000 of capital in exchange for a share of the profits. A would-be investor consulted Babbage on the soundness of the scheme, offering to stake up to £10,000 if Babbage confirmed its merits. Babbage analysed the proposal and, needless to say, found it fallacious. The analysis stimulated him into doing some original work on probability and led to a heavily mathematical paper on games of chance.

He wrote on electricity and magnetism, astronomical instruments, mathematics and diving bells. He had been lowered underwater in a cast-iron vessel during a visit to Plymouth in 1818, though the diving apparatus of the time was primitive and hazardous. Oblivious of the risks to his personal safety, he monitored his own physiological responses to pressure and temperature, and published a long article in 1826 in *Encyclopedia Metropolitana* in which he analysed oxygen intake, ballasting, payloads and propulsion.

In 1826 he published a book on life assurance based on his brief and doomed actuarial experience with a start-up firm, the Protector Life Assurance Company. Benjamin, his father the banker, was well pleased. The book is remarkable in many respects, not least because it was intended for the general public and those unfamiliar with the booming market for assurance. Babbage, principled as ever, was indignant at the 'disgraceful practices' he uncovered during his brief spell in the business. The book, entitled *A Comparative view of the Various Institutions for the Assurance of Lives*, was meant as a handbook, a vade-mecum, for those who might take on trust the supposed benefits of assurance cover and who were therefore vulnerable to exploitation through contractual small print, undeclared penalties, and concealed disbenefits. The spirit of the book is one that recurs in his writings – Babbage as the self-elected defender of the unwary by exposing scam, graft or infelicitous misrepresentation. It is, in short, an early work on consumer protection.

His interest in tables and table-making remained as avid as ever. He was not only a collector and connoisseur of tables but a fastidious analyst of tabular errors. His private collection totalled some three hundred volumes, many of them extremely rare, and was among the most comprehensive sets in the country. Because of the cost, tedium and susceptibility to human error, new editions of commonly used tables were rarely computed from scratch. Existing tables with a reputation for accuracy were used as the starting point. These were checked, extended, reduced or revised for new editions. In 1827 Babbage published his own table of logarithms of the numbers 1 to 108,000. The process by which he verified the integrity of the final version gives some insight not only into Babbage's capacity for meticulous detail, but into the lengths one had to go to achieve any improvement on the existing state of affairs.

He had the seven-figure tables by the Frenchman François Callet re-typeset from the printed version. To ensure that the last digits of the newly set tables were correct, he checked the printed results against Georg von Vega's ten-figure tables and made appropriate corrections. In all, the proofs were checked no fewer than nine times against various sources and any discrepancies corrected. His tables achieved a reputation for reliability, though a hundred years later it emerged that the values for numbers greater than 100,000 suffer from checking errors – they were proof-read in a tent in Ireland during a storm.

Babbage printed sample results using differently coloured inks on variously coloured paper. He argued that the optimal combination for reading by candlelight would be different from that appropriate to daytime reading. He took the ergonomics of reading tables to extraordinary lengths, producing samples using every permutation of ink colour and backing paper including – bizarrely it seems – green ink on three shades of green paper. He was a mathematician and a logician, and a permutation is a permutation. Perhaps it was done tongue in cheek, though anyone inspecting the surviving proofs will find that the combination does not look as daft as it sounds.

His investigation of tabular errors led to a short paper read at an Astronomical Society meeting in March 1827 which revealed the astonishing detail with which he analysed errors.

He found that there were six errors common to nearly all tables, which he attributed to the universal practice of copying. (His own tables and those of Von Vega and Adriaan Vlacq were exempt.) The first two errors he attributed to mistakes in typesetting. The third and fourth he proposed arose when the two pieces of type bearing '4' and '8' fell out in the course of printing and were swapped over when replaced. (The stickiness of printer's ink sometimes pulled type out of the frame during inking or printing.) The fifth and sixth errors he laid at the doors of either the printer or the computer, and he gave possible reasons for both.

During the preparation of his logarithm tables Babbage took Georgiana to Paris – a busman's holiday, it seems, because he used the occasion to check his proofs against the great cadastral tables of de Prony that had so influenced him earlier. In Paris the couple enjoyed the company of the savants, intellectuals and the social elite of the French capital with whom they dined in undoubted style. The image of the happy and charming Charles arm in arm with the devoted Georgiana dressed in their finery, strolling and chatting in the wide boulevards of Paris, is one to hold on to. The idyll of their marriage and of Babbage's personal life was about to be shattered.

In February 1827 his father, Benjamin, died. The prospect of redeeming himself in his father's eyes was now gone, and his sorrow despite the friction between them can only be imagined. Georgiana was a devoted and loving wife, and from what little correspondence survives she comes across as a caring and charitable soul. Yet on Benjamin's death she wrote to Herschel that 'to feign sorrow on so happy a release would in *me* be an hypocrisy' – an uncharacteristically harsh reaction. She appeared to resent Benjamin's denigration of Charles's talents and was pained at the way this had undermined her husband's self-confidence. Herschel had written asking whether Charles wished to be considered for a chair of mathematics at Oxford. In her reply Georgiana wrote:

> One great inducement in endeavouring to procure an addition to our income was his father's always *despairing* his abilities . . . and saying C's abilities would never procure him anything. This made dear C feel more keenly the fruitlessness of his endeavours. This trial is now past.

On his father's death Charles inherited an estate valued at about £100,000. He was now independently wealthy and well able to keep his family and, with prudence, finance a life in science.

The second blow struck in July 1827 when their second son, Charles, died. This was not the first time the couple mourned a lost child. Infant mortality was high, and only four of eight Babbage offspring survived childhood. Following close on his father's death, the loss of young Charles was a heavy burden to bear. Worse was to follow. Georgiana fell ill. At the beginning of August they travelled with the children in great agitation to Georgiana's sister near Worcester. In October, in the final tragedy of that year, Georgiana died, presumably in childbirth, as well as a newborn son.

Babbage was inconsolable. He sought solace with the Herschels in Slough, though his state of mind remained a source of anxiety to his mother who wrote to Herschel: '. . . you give me great comfort in respect to my son's bodily health. I cannot expect the mind's composure will make hasty advance. His love was too strong and the dear object of it too deserving . . .'. Herschel bundled Babbage off to Ireland during September hoping to distract his friend with work. But science was a feeble balm. Babbage was shattered and close to breakdown. On the advice of those close to him he prepared for an extended tour of the Continent to recuperate. Hastily he arranged his affairs. He secured permission from the Government to suspend his responsibilities to the Engine and leave the country. The project had so far cost £3,475. Once the first tranche of government money was spent, Babbage bankrolled the engine construction using his own funds with the intention of reclaiming the expenses from the Treasury. While the accounting was exact, he was casual about how large a deficit he would rack up before making a claim. He was already out of pocket, but there was neither time nor concern to resolve the arrears now. He entrusted the project to Herschel and left £1,000 of his own as an advance on Clement's charges, as well as drawings for the continuation of Clement's work.

Dugald and Henry, the two younger children, were cared for in the household of their late mother's sister, Harriet Isaac, while Herschel and young Georgiana stayed in the family home in Devonshire Street in the care of Babbage's mother. With these makeshift arrangements in place he set off in October 1827, taking one of his workmen, Richard Wright, as a travelling companion.

For over a year Babbage and Wright toured the cities and towns of Europe, journeying through Holland, Belgium, Germany and Italy. The route took in Louvain, Liege, Maastricht, Aachen, Frankfurt, Vienna, Munich, Innsbruck, Verona, Padua, Venice, Parma, Reggio, Bologna, Florence, Rome, Naples, Trieste, Laybach, Gratz, Berlin, Prague and Dresden. Babbage recounts some of his experiences in *Passages from the Life of a Philosopher*, a rambling anecdotal work written when he was in his early seventies, over three decades after his travels. Although *Passages* is an autobiography of sorts, Babbage reveals practically nothing of his emotions or of the state of mind he was in when he set off from England. Nowhere does he mention Georgiana, and there is only one clue to the bleakness of the circumstances of his travels. In Florence he met the Grand Duke of Tuscany, Leopold II. When he visited Turin some dozen or so years later he was granted several audiences with the King of Sardinia, who was fascinated by this imaginative and knowledgeable foreigner. During the later visit Babbage made a special effort to see the Queen of Sardinia, the sister of the Grand Duke, as a gesture of gratitude for her brother's consolation and comfort 'when under severe affliction from the loss of a large portion of my family'. This is the only reference in *Passages* to Georgiana and the trauma of 1827, and an oblique one at that.

Passages is a lively compilation of recollections, amusing and instructive episodes, hobby horses, score-settling, self-justification, parody, puff and charm. By the time he came to relate episodes from his travels in 1827–8, the edge of the pain that had prompted his Continental tour had dulled, and he invested his escapades with a vigour and relish that he could not have felt at the time.

Master and man visited tourist attractions, factories, workshops and universities and, as a distinguished man of science, Babbage dined with the intellectual and social elite of Europe. By the time he reached Vienna he was a seasoned traveller and had a four-wheeled 'camper' built to his own design. It was equipped with all manner of travelling conveniences, sliding drawers, pockets and pouches for the paraphernalia of touring (books, money, telescopes), space for dress coats, sleeping accommodation and rudimentary cooking facilities. (The camper can be seen as an antecedent of the Volkswagen 'Combi', and pre-echoing the practice among travellers 150 years later, Babbage would sell the vehicle on the return leg in Holland – for half the purchase price.)

While travelling, Babbage became anxious to confirm his original understanding that the expenses he incurred on the

Engine would be reimbursed by the Treasury. He wrote from the Continent to his brother-in-law, Wolryche Whitmore, urging him to seek confirmation from John Robinson (recently ennobled to Viscount Goderich) that Babbage was correct in his understanding of the Government's obligations. Robinson was evasive. He refused to be pinned down, and the matter was left unresolved.

In Rome, Babbage happened to pick up a copy of *Galignani's Messenger*, a kind of European chronicle published in English. Much like the modern *Herald Tribune* today, it featured a digest of reprints and articles from national newspapers. *Galignani's* was widely read throughout Europe and favoured by English-speaking travellers. Babbage happened on an entry announcing his appointment to the Lucasian Chair of Mathematics at Cambridge: 'Yesterday the bells of St. Mary rang on the election of Mr. Babbage as Lucasian Professor of Mathematics.'

The Lucasian Chair had once been held by Isaac Newton, and was among the most prestigious mathematics appointments to be had. The salary was modest – under £100 – and through a loophole in the regulations the incumbent professor did not need to reside in Cambridge, nor indeed give lectures. Babbage had applied for the post in 1826. He was then concerned that the duties of the appointment would interfere with his commitments to the Difference Engine, and he took pains to confirm that the obligations of the post were neither onerous nor time consuming. He recalled that when he read the news he had mixed feelings. Before he left England he had begun to sense the unexplored potential of the Engine and was still anxious that the duties of the Chair, though light, would be a distraction. He drafted a dignified refusal but was persuaded by friends to accept, which he duly did.

His 'occupation' of the chair, which he held from 1828 to 1839, is a figure of speech in more respects than the obvious. To the resentment of the dons he never resided in Cambridge and gave no lectures, though he conscientiously fulfilled his obligations as an examiner for the coveted Smith's Prize in mathematics. He was deeply gratified by the appointment and later observed without sarcasm that it was 'the only honour I ever received in my own country'. His pleasure was doubtless doubled by his recollection of the debacle of his Acts and for not competing for mathematics honours in the final examinations of the tripos.

Back in England there were ominous rumblings. It was five years since the announcement of government financing of the Engine project. With nothing credible to show in the way of a physical machine, members of the public began to question the proprieties with which the large sums of public money had been expended. While Babbage was abroad an article appeared in the *Record* calling on Babbage and his friends to account for the funds. It insinuated that the project had failed, that Babbage was guilty of concealment and, most damaging of all, that he was profiting personally from the venture.

Loyal Herschel leapt to his friend's defence. Herschel had been entrusted with the supervision of the project during Babbage's absence and had first-hand knowledge of progress and expenditure. In a letter to *The Times* he attested from his 'certain knowledge' that all public moneys received for the engine had been expended solely on its proper object and no other, and that, far from profiting personally, Babbage had made up the frequent deficits from his own pocket. He denied that the project had foundered:

> The work, meanwhile, continues in active and steady progress, but such is its extent, such the variety of

> mechanical movements to be contrived and executed and such the elaborate perfection of workmanship which has been found necessary to bestow on all its parts . . . that a very long time must yet elapse, and a very heavy further expense be incurred, before it can be completed: but no suspicion of a failure has yet arisen. On the contrary, every mechanical difficulty has been completely overcome, nor has an obstacle occurred in the slightest degree calculated to raise a doubt as to its ultimate success.

Babbage was hurt by any suggestion of mistrust that impugned his own conduct. His sense of injury was aggravated by the personal and financial sacrifices he had made in the service of what he regarded as the fulfilment of the Government's wishes and for the greater good of science, a mission for which he had not asked to be paid. The wounds were deep and the grievance bitter.

Babbage returned to England at the end of November 1828 after just over a year away. On his return he tried to resolve the long-standing uncertainty about the nature and extent of the Government's commitment to the Engine. There was no Treasury record of the original meeting with the Chancellor, John Robinson. There is no question that Babbage left the meeting convinced in his own mind that the Government had committed to fund the engine to completion. But no budget was agreed. There was no ceiling on expenditure and no delivery date. On 6 December 1828 Babbage made a direct appeal to the Duke of Wellington, then Prime Minister, in the form of a long letter. He pointed out that some £6,000 had already been expended, of which all but £1,500 he had paid himself. He also hinted at the stress of the financial uncertainties: 'the additional anxiety thus created would be highly unfavourable to that state of mind most fitted for the performance of this and other scientific duties'.

Wellington referred the matter to the Royal Society, which duly appointed a committee under the chairmanship of none other than faithful Herschel. The committee's brief was to advise whether the progress on the machine justified the continued belief that it would accomplish its original object, the automatic tabulation and printing of mathematical tables. The recommendations of the committee were again favourable. They commended progress and defended Babbage's insistence on nothing but the highest precision throughout. The report, delivered in February 1829, concluded with the hope that while Babbage was so earnestly engaged 'he may be relieved, as much as possible, from all other sources of anxiety' – a coded bid for more money. This duly came in April 1829 in the form of a formal Treasury minute directing payment of a further £1,500, a sum which still left Babbage out of pocket. But the timing was fortunate: Babbage was in the process of moving to 1 Dorset Street, off Manchester Square, the former home of his friend William Hyde Wollaston, who had recently died. Babbage's mother, Elizabeth, continued to occupy the house he had shared with Georgiana in Devonshire Street, a few blocks away. 1 Dorset Street remained his address for the rest of his life.

Babbage was far from recovered from the tragedy that had prompted his travels. He was ill and depressed, and in search of relief he lost himself in work. In March 1829 he wrote to the Reverend Edward Smedley:

> I have suffered so severely in health that however much I desire to be active, all my friends, and more especially my *medical* ones, urge me to lay aside my

> pursuits and to set my mind at rest or asleep. The alternative . . . seems to be the *long sleep*. The intensive occupation I have used as a remedy has had its effect. I have lived through two years and I may live two or twenty more, but the medicine has produced a disorder, and if I am to finish the machine I must cure the new complaint. I have therefore decided to give up everything but that.

As if despondency and ill-health were not enough, his relationship with Clement began to show cracks. The system of ad hoc payments was clearly unsatisfactory. To put the arrangements on a more business-like footing Babbage asked two well-known engineers, Bryan Donkin and George Rennie, both of whom had served on two of the Royal Society's Engine committees, to examine Clement's bills. He also asked them to raise with Clement a series of fundamental issues that should have been resolved much earlier. He sought to clarify who owned the tools, patterns and drawings; at whose risk were the machines and plans; and how the assets of the project were to be secured from Clement's creditors in the event that Clement ran into financial difficulties. He further requested an undertaking from Clement that he would not make copies of the Engine without written permission from Babbage. The issue was not one of mistrust or suspicion – Babbage thought highly of Clement and believed him fully deserving of proper reward. But things got messy. The upshot was as clear as it was disturbing: the tools were Clement's, the patterns and plans Babbage's, and each should insure his own property. And, no, Mr Clement would not agree to secure permission from Babbage to make copies of the Engine.

Clement was within his rights to claim ownership of the tools. The practice was established to cover hand tools for journeymen and craftsmen. There was no special provision that would exempt the machines, jigs and specialised apparatus developed at the project's expense by Clement himself or in collaboration with Babbage. Clement's refusal to seek permission to make copies of the Engine was a demoralising blow, but he did agree to compile details of his charges. This was certainly in his interest: he had been paid £3,260 but was still owed over £2,000. But, either through obduracy or misunderstanding, he failed to supply the financial information. Babbage in turn was determined to make no further payments until the accounts had been vetted to his satisfaction, especially as the financial obligations of the Government were still unclear. Both were stubborn, determined, and convinced of the justice of their own position. Stalemate. Clement halted work on the engine in May 1829. The stand-off was resolved through arbitration. Clement's bills were found to be fair and correct, and his account was settled in May 1830. At long last he resumed work. The Engine had lain dormant for over a year.

The project continued to be dogged by fitful finances. The insecurity of funding and the protracted stand-off with Clement were vexing. Although Babbage was beleaguered, he was far from idle. The movement to reform the management of English science was gaining momentum, and Babbage was a prominent activist campaigning and lobbying behind the scenes. In May 1830, about the time Clement restarted work on the Difference Engine, Babbage rocked the scientific establishment to its core. His book *Reflections on the Decline of Science in England, and on Some of its Causes* was published and caused a spectacularly bitter furore.

The movement to reform the Royal Society had been active for some years, with Babbage, Herschel, Wollaston and others of Babbage's circle among its ringleaders. The group was spurred on by its conviction, shared by others, that since the glory days of Newton, science in England was in general decline, but more specifically that the Royal Society was defective in its superintendence of scientific life. Babbage's 'declinist' views were given a rousing boost by his experiences on the Continent. He was impressed by the social and political prestige of scientists, particularly in France and Prussia, and envious of the patronage they enjoyed from both royalty and the state. He was equally impressed by the high standards demanded for membership of scientific academies and societies, and the efficiency of their administration.

The 'declinists' railed at the neglect of science by the English aristocracy and the state, and lamented the indifference to the welfare of science's practitioners. Babbage cited the rarity of civil honours to scientists as one indicator of this indifference, and bemoaned the scientific ignorance displayed by the government and the emphasis on classics in educational curricula. There is no doubt that the professionalisation of science was significantly more advanced in France and Prussia than in England, and Babbage, champion of the just cause, was fired by the discrepant fortunes of his colleagues and friends across the Channel.

To the 'declinists', the Royal Society was a self-serving coterie of amateurs unfit to advance the interests of science. The Society was the wretched target of Babbage's sarcastic salvos, and the book reads for the most part as a chronicle of procedural misconduct by its officers. He spares his quarry the courtesies of gentle chiding – for him, nothing as subtle as a quiet word in the odd ear. Instead he unleashes a broadside of outrage and insult in which he attacks the personal and professional probity of his victims. As with much of Babbage's campaigning writing, he argues more to protest than to persuade. He identifies his villains by name, accusing various officers of forging minutes and rigging the award of medals, and impugning no less a figure than Sir Humphry Davy, the previous president, for misappropriating funds for personal profit. He goes on to accuse its committees of maladministration and of failing to detect scientific fraud in its own publications. There were times when Babbage behaved as though being right entitled him to be rude, and the strength of his convictions tended to make him insensitive to the effect of his actions on others. These caustic *ad hominem* public attacks were a shocking breach of the conventions of the day, according to which displeasure was more likely to be settled by intrigue, collusion, brandy and cigars. Babbage's public outbursts have earned him the sobriquet 'Irascible Genius' – the title of the first published biography of Babbage, by Maboth Mosely, which appeared in 1964. This caricature of Babbage's public persona offers a sharper and more credible historical image than many of the scholarly studies published since.

Although Babbage's views were shared by others, he was alone in the vehemence of their public expression. His friends advised him against going to press. Herschel, who was shown a pre-publication draft, told him:

> If I were near you and could do it without hurting you and thought you would not return it with interest I would give you a good slap in the face. Now are not you ashamed of yourself to keep up your old growl on the score of having expended time and money in accomplishing a great and worthy object which will hand you down with well-earned fame to posterity? Who the deuce ever did anything worth naming without sacrifice?

Herschel saw Babbage's exasperation with the Engine project – the tardy progress, the financial crises, the confrontation with Clement – as the root cause of his immoderate rage. Certainly the intensity of the outrage seems at times disproportionate to the apparent veniality of the offence. The frustrations with the Engine surely played their part. But the bitterness of the attacks is perhaps an expression of Babbage's pain, anger and grief at the devastating loss of Georgiana, two of his children and his father in the space of ten months, and the book may have been a public vehicle for personal despair.

Decline gave a decisive boost to the movement to reform the institutions of science, forcing the issues of organisational conduct into open discussion. While the blast was seen by many of Babbage's contemporaries as an ill-considered affront, he was also privately hurrahed by those delighted to have acquired a champion without having to put themselves at risk. Herschel, for one, shared Babbage's declinist views, but he was more circumspect in his advocacy. In time, he became a fine ambassador for English science. Babbage, on the other hand, was cast in the role of *enfant terrible*. We have seen how prudence and diplomacy had served Herschel as well at Cambridge as public protest had served Babbage badly. The episode of *Decline* mirrored the contrasting fate of the two gifted friends.

Babbage had at a stroke alienated the selfsame people whose support he needed, and at the same time soured his relationship with the pre-eminent scientific body whose committees had three times recommended government support for his Engine. However justified his concerns for the welfare of science in England may have been, public insult was not the best way to win friends and influence people. Expulsion from the Society was whispered about behind closed doors. When the furore over the book had died down, this move of censure faded, but Babbage's position as an insider was irreparably compromised. He later wrote to the Duke of Somerset that since the publication of *Decline* he had 'never attended the Royal Society nor even indirectly taken any part in its affairs'. As an instrument of reform, *Decline* served its purpose well. As a career move it was a disaster.

With Clement back at the bench, Babbage estimated that he would complete the Engine in about three years – ten years since the heady first days of government support. Work progressed steadily, though not uneventfully, for the next two years. With finished parts accumulating at Clement's works, Babbage could now foresee the time when he could at last begin to assemble the great machine. The Engine would be massive and heavy, and not easily moved. He surveyed his neighbourhood, looking for a site conveniently close to his house to assemble and eventually operate the beast. A site was found close by. Estimated cost: £2,250. By now everyone knew the routine – application to the Treasury, referral to the Royal Society, favourable outcome. And so it came to pass. In March 1831 the Royal Society endorsed the proposal for new premises, and Brunel estimated that the Treasury should allow for the shockingly large sum of an additional £12,000 to complete the Engine.

As the new fireproof buildings took shape, Babbage began to plan their use. He intended that Clement and his family would live on the premises and that provision should be made for accommodating 'clients' – staff associated with the production of the *Nautical Almanac*, for example. To comply with Babbage's wishes Clement would have to split his establishment, and in July 1832 he insisted on being compensated to the tune of £660 a year for loss of business and inconvenience.

With the new building in prospect Babbage instructed

Clement to assemble a small section of the Engine from the already completed parts. It is possible that Babbage wanted to rehearse the assembly process and verify the mechanical and logical operation of the device. It is also possible that the issue of credibility played its part. After nearly a decade of design, development and swingeing costs there was little to show, and perhaps he wished through some tangible evidence to demonstrate that the grand venture was not a piece of unfounded optimism or intellectual grandiosity. A working section of the Engine would help to silence his detractors, especially those who had decried the lavish expenditure of public moneys.

The demonstration piece was delivered to Babbage's house in Dorset Street in December 1832 and was placed on display in his drawing room. It stood about two and a half feet high, two feet wide and two feet deep – a weighty bronze and steel embodiment of solidity and precision. It consisted of three columns, each with six engraved figure wheels, representing about one-seventh of the complete calculating section. It incorporated the essential calculating mechanism which is repeated time and again in the full design, though no printing section was made for it.

The Treasury rejected Clement's demand for compensation as 'unreasonable and inadmissible'. From here on it is difficult to untangle the precise sequence in the downward spiral to the final impasse. Clement submitted a bill. Babbage insisted that Clement withdraw or revise his demand for compensation. Clement refused. Babbage refused to advance any further money from his own pocket and invited Clement to submit his bill to the Treasury. Clement threatened to lay off his men if he was not paid. Babbage was unyielding. Clement put his men on notice. Babbage was unmoved. At the end of March 1833 Clement's men were fired, Joseph Whitworth included.

The breach with Clement was final. Work on the Engine stopped, never to be resumed.

The return of the drawings and finished parts and the final settlement of bills was a gruelling business involving intermediaries and mistrustful checking of lists and inventory. Arbitrators and mediators were used in a grudging three-way negotiation between a grumpy Babbage, an intransigent Clement and the long-suffering Treasury. Babbage wrote that he was 'almost worn out with disgust and annoyance at the whole affair'.

By the time of Clement's last payment in August 1834 the Difference Engine project had cost the Treasury a total of £17,478 14s 10d, of which £2,190 13s 6d was for the fireproof buildings, and the rest for manufacture and development. This was a staggering sum of money. A brand-new steam locomotive, the *John Bull*, built by Robert Stephenson for shipment to the United States in 1831, cost all of £784 7s.

By the time Clement downed tools he had made 12,000 of the requisite 25,000 parts. Most of the parts for the calculating section of the Engine were complete but unassembled. The parts for the printing apparatus were scarcely begun. Once the loose parts were returned to Babbage, some were recycled for experiments and testpieces by both Babbage and later his son, Henry Prevost, who also assembled sets of the original Engine parts into small demonstration pieces after his father's death. The fate of the rest, Clement's best and some of the finest examples of precision engineering of the day, was to be the melting pot, or sale by weight for scrap.

Although work had stopped, Clement was reluctant to give up the lucrative and prestigious contract, and he proposed that he continue to manufacture parts, deliver them to Babbage's house, and assemble the machine in Babbage's new workshops. But the spirit of collaboration between the two men was broken and there was nothing to be salvaged from the close association they had enjoyed in years gone by. Perhaps Clement dawdled. Perhaps he hoped that the longer the winding-up process took, and the more of a meal he made of it, the better his chances of a continuation of sorts. Perhaps he was just an honourable stickler for what he felt was his due. Finally, on 15 July 1834, Babbage took delivery of the coveted drawings and thousands of loose parts already made. It had taken fully sixteen months for the wretched affair to be settled.

It is tempting to demonise Clement, and see him as a saboteur of a great historic enterprise and Babbage as the thwarted victim of Clement's stubbornness and even greed. One of Clement's workmen, Charles Godfrey Jarvis, had his suspicions about Clement's motives. Jarvis resigned from Clement's employ and wrote to Babbage in February 1831:

> It should be borne in mind that the *inventor* of a machine and the *maker* of it have two distinct ends to attain. The object of the first is to make the machine as complete as possible. The object of the second – and we have no right to expect he will be influenced by any other feeling – is *profit*: to gain as much as possible by making the machine; and it is his interest to make it as complicated as he is permitted to make it. I am fully aware how far these observations may do me injury. But they are made, Sir, whether well or ill judged, for your good.

It is difficult to know whether the allegation that Clement was over-complicating the machine out of self-interest are the words of an aggrieved ex-employee rubbishing his former boss and ingratiating himself into Babbage's favour in the process.

Jarvis blackens Clement further in a letter to Babbage written in August 1833, after Clement's final breach with the Engine project:

> To a man who although inactive and unenterprising loves money, it must be very agreeable to construct a newly invented machine, the cost of the parts of which cannot be taxed . . . and . . . to charge for the time expended . . . without the possibility of the useful employment of that time being disputed, and to doze over the construction year after year for the purpose of making one thing after another . . .

The allegation is clear: that Clement was making a meal of the job to prolong a lucrative open-ended contract. There is evidence to support this. Measurements made at the Science Museum in the 1980s on the portion of the Engine completed by Clement show that non-critical parts, such as spacing pillars, are machined to the same high tolerances as operationally critical parts. If Jarvis's motive was to discredit Clement and secure employment with Babbage, then the ploy paid off. He was later hired by Babbage at 2s 6d an hour and served him well for over a decade, producing 'all the beautiful drawings for the Analytical Engine'.

Clement won acclaim for his machines. He was awarded the Society of Arts gold medal in 1827 for lathe improvements, and the next year the silver medal for his 'self-adjusting double-driving centre chuck'. His most significant influence was not officially rewarded – his attempts to standardise screw threads and the influence this had on Whitworth, who carried the programme through with spectacular success. Clement owes much to Babbage. Most of the machines for which he was applauded were built while he was working on the Engine and largely at Babbage's expense. Babbage was aware of this. Two years before his death he wrote:

> I have heard at different times from men I had employed in former years that amongst their own class it was frequently said that: Mr. Babbage made Clement. Clement made Whitworth. Whitworth made the tools. When I first employed Clement he possessed one lathe (a very good one) and his workshop was in a small front kitchen. When I ceased to employ him he valued his tools at several thousand pounds and he had converted a large chapel into workshops.

Clement had set up his own workshop with premises south of the river in 1817, about seven years before he was hired by Babbage. Babbage wrote this account forty-five years later, and the depiction of the modesty of Clement's circumstances may be wishful. But there is no question that Clement's rise was the result of his own talents as well as the financial and technical opportunities presented by Babbage and his Engine.

The Engine project had attracted public attention and may well have made its way into contemporary fiction. Charles Dickens's novel *Little Dorrit* introduces the Circumlocution Office, thought to be a satirisation of the Treasury and its dealings with Babbage over the Engine. There is also a character called Daniel Doyce, a worthy inventor-engineer whose attempts to secure government support for his great invention are thwarted by the labyrinthine obstacles of bureaucracy. This is Dickens's early description of Doyce:

> The ingenious culprit was a man of great modesty and good sense; and, though a plain man, had been too

> much accustomed to combine what was original and daring in conception with what was patient and minute execution, to be by any means an ordinary man.

Doyce, like many of Dickens's protagonists, is portrayed as the wholesome hero and can be seen as a composite of the best of Babbage and Clement – inventive genius on the one hand, and practicality on the other – with the less creditable features of each fictionalised out – prickliness in the case of Babbage, and the tendency to overcharge in the case of Clement. Despairing of support or recognition in England, Doyce goes abroad where for a while he flourishes – a reflection again, perhaps, of Babbage's admiration for all things foreign. The parallels in *Little Dorrit* are not fanciful. Dickens and Babbage were part of the same social set and met frequently at fashionable parties and dinners.

Clement's quitting the project in 1833 was a severe blow. But Babbage had not yet given up hope that his Engine might still be completed.

Chapter 4

MIRACLES AND MACHINES

All were eager to go to his glorious soirées.

Harriet Martineau, 1832

Babbage slowly emerged from the darkness of Georgiana's death. He socialised more and began to host parties. His soirées were sparkling events in the London social calendar. On Saturday evenings, No. 1 Dorset Street was the hub of social, literary and intellectual life in London with 200 to 300 celebrities, civil dignitaries, authors, actors, scientists, bishops, bankers, politicians, industrialists and socialites converging for gossip and intrigue, and of course the latest in science, literature, philosophy and art. Babbage had his own criteria for selecting who would attend. A contemporary wrote that 'One of three qualifications were [*sic*] necessary for those who sought to be invited – intellect, beauty, or rank – without one of these, you may be as rich as Croesus – and yet be told, you cannot enter here.'

Babbage's teenage sons, Henry and Dugald, sometimes

attended. Henry later wrote that 'there were all sorts of people there. I saw seven Bishops there one evening and the best of almost every class.' The greats of the day went to Babbage's: Charles Dickens, Brunel, the famous actor William Macready, Charles Darwin, pioneer of photography Fox Talbot – the list goes on. Objects of scientific and artistic curiosity were exhibited for the entertainment and instruction of guests – the latest results of new photographic processes, specimens, novel devices, automata – and hopeful inventors petitioned members of the privileged set to display their inventions, contrivances and artistic works to the toffs of culture. 'All were eager to go to his glorious soirées,' wrote Harriet Martineau, a literary figure of the time famed for her popular novellas, and invitations were highly prized.

Not everyone was preoccupied with the advancement of mind. The geologist Charles Lyell pressed Babbage to invite Colonel Codrington's wife, whom Lyell had heard was 'very pretty'. Lyell also urged Charles Darwin, returned from his five-year adventure on the *Beagle* in 1836, to attend Babbage's where he would meet the fashionable intelligentsia and, more to the point, beautiful women. One Woronzow Greig asked for an Engine demonstration to impress two lady friends, a Miss Parker and a Miss Sandbath from Liverpool, 'very young and very pretty'. Babbage became a sought-after dinner guest. He was a celebrity, an engaging raconteur, full of wit and exuberant invention. To be able to say 'Mr. Babbage is coming to dinner' was the pleasure and delight of any hostess.

In the early 1830s there was much to excite the chattering classes. There was a general sense of turmoil and impending change intensified by the debates that raged over the reform of the voting system and the terms of the great Reform Bill of 1832. Babbage was politically active at this time. He stood as a Liberal candidate for the Borough of Finsbury in the general election of 1832, and again in the by-election of 1834. During the campaign he was barracked by questions about the financial proprieties of the Engine project – the touchiest of his sensitivities. He was defeated in both elections, and abandoned further political ambition. With his political boiler well stoked, he privately printed a pamphlet entitled *A Word to the Wise*, a vituperative attack on the system of hereditary privilege and titles which he wanted abolished in favour of life peerages. In it he attacked the 'insatiable desire for political power' of the aristocracy and the 'disgraceful corruption' to which it leads. When Babbage was steamed up there were no half-measures.

Science was flourishing, despite the arguable competence of the Royal Society in the administration of scientific life. The early Victorian elite had inherited the notion that all that was known or knowable could be mastered by an individual. Being a know-all was not yet a presumption but a legitimate aspiration. As scientific activity increased, science began to fragment into specialisations, and the realisation that the entirety of scientific knowledge could no longer be marshalled by a single practitioner began to surface. New scientific societies began to press for recognition, and the pressure culminated in the founding of the British Association for the Advancement of Science in 1831.

The Association provided a broad organisational umbrella for the new diversity. Its separate sections, committees and subcommittees, covered the gamut of scientific knowledge – mathematical and physical sciences, chemistry, mineralogy, geology, geography, zoology, botany and the mechanical arts. To these were added electricity, magnetism, physiology and anatomy. The breadth of its compass, the massiveness of its annual conventions and the obsessive detail of its proceedings prompted Charles Dickens to parody the new organisation as the 'Mudfog Association for the Advancement of Everything'. Babbage was in the thick of it. When it came to reform and mucking in with a renegade group, steam engines could not drag him away. He was active in the founding of the new organisation and from 1832 served for six years as one of its trustees. He was also responsible for instituting the Statistical Section of the Association, which in turn led to the founding of the Statistical Society of London in 1834.

Science was increasingly topical. Mesmerism was taken seriously and its supposed therapeutic benefits were eagerly sought. Phrenology, which purported to reveal character and ability from the shape of the head, was used in political arguments against hereditary privilege, patronage and career preferment. If talents were evident from bumps on the head, then those of high birth or with social connections had a new test to pass, and superior ability could not be assumed to be a natural consequence of class. Merit and privilege were in the process of being uncoupled. Early evolutionary theory, with its notions of competition between species, was being used as a moral justification for capitalism – nature licensing a competitive free-for-all. For the first time the voice of science was as much part of the debate as were the traditional voices of the Church, vested interest and privilege.

The established order was being challenged by a cheeky new arrival. As science shouldered its way into public and political life, the fault lines between science and religion became stressed as they competed for control over the accepted view of the world. Natural theology taught that the Earth was just a few thousand years old. But geology was producing specimens demonstrably older than this, mounting an uncomfortable challenge to the secure orthodoxies of old. Pre-Darwinian evolutionary theories were beginning to challenge the Creationist accounts of the Book of Genesis. If we evolved from primeval slime, where do we fit the teaching that God created Heaven and Earth and all creatures therein?

Miracles posed particular difficulties for rational science. By definition, miracles are events without evident cause. Religion taught that God was the causal agent of all events, so no further explanation was necessary. But the central tenet of rational science is that events have physical rather than divine cause, and physical cause was the main plank in science's campaign for the monopoly of explanation. Miraculous phenomena were a real headache for the rationalists. But miracles posed no such problem for religion: they were not only manifest proof of the omnipotence of God, but very good PR. If science was not to lose out it would have to engage with miracles and solve the intolerable paradox of a physical event without evident cause.

There was a growing fashion for statistics. Babbage used his knowledge of probability theory to engage with the issue of miracles. He interpreted the word 'miraculous' to mean 'improbable', and in so doing brought the miraculous within the grasp of statistics and the mathematics of probability. We find Babbage computing the numerical probability of resurrection from the dead using probability calculus. He did this by estimating the total number of people that had ever lived since Creation, and divided that into the number of times there had been a witnessed account of someone being brought back from the dead (one). He came up with the odds of resurrection as 200,000 million to one against – all without a flicker of absurdity.

Babbage's serious-minded abstractions were a source of affectionate amusement to his friends. The geologist Charles Lyell wrote in January 1832:

> We have had great fun in laughing at Babbage, who unconsciously jokes and reasons in high mathematics,

> talks of the 'algebraic equation' of such a one's character in regard to the truth of his stories . . . I remarked that the paint on Fitton's house would not stand, on which Babbage said 'no: painting a house outside is calculating by the index of minus one', or some such phrase, which made us stare; so that he said gravely by way of explanation, 'That is to say, I am assuming revenue to be a function.' All this without pedantry . . .

Babbage wrote to the poet Tennyson, putting him right on the arithmetic of population growth:

> Sir: In your otherwise beautiful poem 'The Vision of Sin' there is a verse which reads – 'Every moment dies a man, Every moment one is born.' It must be manifest that if this were true, the population of the world would be at a standstill . . . I would suggest that in the next edition of your poem you have it read – 'Every moment dies a man, Every moment 1$\frac{1}{16}$ is born.' . . . The actual figure is so long I cannot get it onto a line, but I believe the figure 1$\frac{1}{16}$ will be sufficiently accurate for poetry.

Babbage's involvement with the unexpected did not stop at estimating the chances of returning from the dead. He had a more developed theory of miracles, and he used the portion of his Difference Engine in a dramatic demonstration of his argument. His theory of miracles was a party piece at Saturday soirées where he entertained and instructed the glitterati of London society. Before his guests arrived he would set the machine up so that each time the handle was cranked it performed a simple arithmetical rule – say adding '2' to some number already there.

When the soirée was in full swing he performed his piece. We can picture him in his drawing room with the Engine, surrounded by a cluster of fashionably dressed ladies and gents gathered around the Engine for the intellectual cabaret. The tittering fades to silence as Babbage, the host, takes his place and starts to expound. 'Observe the numbers on these wheels,' he says pointing to the elegantly engraved bronze figure wheels arrayed in neat columns in the machine, 'they are all set to zero.' Guests lean forward, and see that all the wheels on the last column indeed show '0'. Word passes back to those with restricted views – yes, the wheels are set to zero. 'Each time I turn the handle,' he continues, 'the numbers on these wheels will change. Please observe, if you will.' He turns to the machine and cranks the handle. The mechanism operates smoothly and almost silently. 'You see that the number has changed from "0" to "2". Now observe again.' He cranks the handle again and the number changes to '4'. The next repetition advances the number to '6'.

Soon it becomes evident to his guests that the numbers change in accordance with a simple rule – add '2'. He asks them to guess the next number, and makes a show of congratulation when indeed the next result is two more than the previous one. Babbage cranks on. Each time the result increments by '2', and the guests get to know what to expect. He performs repetition after repetition. There is a mechanical inevitability to each outcome and his audience begin to wonder whether Babbage's machine is after all quite banal. But there is a half-smile on his face that keeps them intrigued, and he carries on with the seemingly endless sequence.

After, say, a hundred repetitions, the onlookers witness a remarkable event. Without him altering the settings or interfering with the machine in any way, at the next turn of the handle the result leaps not by the habitual '2' but by a different number, say '117'. Perplexed titters ripple through the throng. Babbage stops cranking the handle. He turns from the machine and addresses his expectant guests. 'You see, to you, the onlookers, this unexpected leap is a violation of law, that is the law of incrementing by "2". But I instructed the machine before I started so that after one hundred repetitions it would add not "2" but "117". So for me the programmer, the discontinuity was not a violation of law but the manifestation of a higher law known to me but not to you. By analogy,' concludes Babbage with a flourish, 'miracles in nature are not violations of natural law, but the manifestation of a higher law, God's law, as yet unknown.' Silence, and then delighted applause.

For Babbage, sudden events in nature, anomalies and unexpected catastrophes were not necessarily the direct result of divine intervention but programmed discontinuities. God was a programmer. (Those in the software industry may be flattered to know this.) In this way Babbage attempted to reconcile the notions of rational order and miraculous events, using his Engine to demonstrate that rules did not always entail uniformity and that one could believe in God and in physical law without conflict.

Control engineering seems also to have been part of God's education. Babbage wrote of the deity revisiting the Earth at some distant time to check for empirical deviations from the grand design which God would then correct. The fact that such deviations may compromise divine perfection did not seem to bother him. For Babbage, empiricism embodied the highest ideal of human knowledge and symbolised a perfection of sorts. So God was clearly an empiricist, and perfection lay in the detection of deviation and its subsequent correction.

Babbage's demonstrations were a sensation and much talked about. But they were more than a party trick. His description of his theory of miracles appeared in a portentously titled philosophical work he published in 1837 called *The Ninth Bridgewater Treatise*. The arguments about lawlike discontinuity were of pressing interest to geologists, who were pondering the causes of earthquakes and abrupt changes in the earth's crust, and to zoologists, who were puzzling over explanations for the sudden appearance of new species after millions of years without change. Geologists and pre-Darwinian evolutionists pored over pre-publication proofs of Babbage's new work. Babbage argued that a rule could continue unchanged for millions of years and then be subject to a pre-programmed discontinuity analogous to the appearance of new species. A machine that could exhibit discontinuous behaviour according to rules provided a physical model of causation that was compelling in ways that philosophical speculation was not. Geologists and evolutionists were buzzing. The portion of the Difference Engine represents an elegant example of the role played by a physical artefact in the history of ideas.

That Babbage had the ability to engage and transfix visitors with his speculations and his Engine we know from eyewitness accounts. Once he got going, he was difficult to stop and he carried his audience with him. One of his visitors, George Ticknor, described a visit to Babbage:

> From church we went, by his special invitation, to see Babbage's calculating machine; and I must say, that during an explanation which lasted between two and three hours, given by himself with great spirit, the wonder at its incomprehensible powers grew upon us every moment. The first thing that struck me was its small size, being only about two feet wide, two feet deep, and two and a half high. The second very striking circumstance was the fact that the inventor

> himself does not profess to know all the powers of the machine; that he has sometimes been quite surprised at some of its capabilities; and that without previous calculation he cannot always tell whether it will, or will not work out a given table. The third was that he can set it to do a certain regular operation, as, for instance, counting 1, 2, 3, 4; and then determine that, at any given number, say the 10,000th, it shall *change* and take a different ratio, like triangular numbers, 1, 3, 6, 9, 12, etc. [*sic*]; and afterwards at any other given point, say, 10,550, change again to another ratio. The whole, of course, seems incomprehensible, without the exercise of volition and thought.

Babbage's ability to fascinate and astound was recalled much later by Lyon Playfair, a scientist and important administrator of science, who missed a lunch appointment:

> Another philosopher whom I frequently visited was Babbage . . . Babbage was full of information which he gave in an attractive way. I once went to breakfast with him at nine o'clock. He explained to me the working of his calculating machine, and afterwards his method of signalling by . . . lights. As I was engaged to lunch at one o'clock, I looked at my watch, which indicated the hour of four. This appeared obviously impossible so I went into the hall to look for the correct time, and to my astonishment that also gave the hour as four. The philosopher had in fact been so fascinating in his descriptions and conversation that neither he nor I had noticed the lapse of time.

The device survives intact, and is on public display at the Science Museum in London. It still works impeccably and calculates, as Babbage intended, without error.

In the complete machine the mechanism was to be powered by an operator turning a large crank handle. The figure wheels would be driven in their deliberate dance by trains of gears and levers derived from this single source. Clement had not yet built the control system of cams, gears, racks and levers to translate the turning of the drive handle into the motions for calculation. So Babbage had a special drive section fitted to the demonstration piece so that the mechanism would calculate automatically when the operator moved a drive handle back and forth.

The demonstration piece occupies a unique place in the history of technology. To begin with, it is one of the finest examples of precision engineering of the time. It serves as a metrological benchmark – a standard of measurement – for what was achievable by the most advanced practices of the day. If you want to know about the limits of achievable precision in the 1820s, then the demonstration piece provides a prime example from which to take measurements. Questions about the limits of repeatability – how exactly parts could be made to resemble each other – can be answered by measuring the near-identical wheels and gears on the machine and analysing the dimensional 'spread'. Metallurgical puzzles about the composition of contemporary materials – gunmetal, brass, cast iron and steel – can be solved by analysis of the metals used for its parts. The stiffness of shafts, the degree of interchangeability of parts and the hand-grained finish are some of the curatorial treasures enshrined in this device.

The relevance of the piece to the history of manufacturing is of interest to specialists. But its main feature elevates it into a category all its own, and secures for it an uncontested place in the history of computation. We have seen that the *idea* of an automatic printing calculator and the use of differences was not new: it had been suggested by Johann Helfrich Müller over thirty years before Babbage's first design. The revolutionary and unique feature of what Babbage referred to as the 'finished portion of the unfinished engine' is that it is completely *automatic*.

We saw earlier that the mechanical calculators of the time required an operator with some degree of mathematical (or at least arithmetical) knowledge to manipulate the dials and sliders in a fixed sequence. Using the ornamental curiosities of Pascal and Leibniz, and the later arithmometer of Thomas de Colmar, calculation progressed only by a succession of manual operations performed by the user. Without the continuous informed intervention of the operator these devices were no more than elaborate scientific showpieces.

Babbage's machine is the first known calculating device to successfully embody mathematical rule in mechanism, and it symbolises the start of the era of automatic computation. Once the figure wheels had been set with the starting values, all you did was turn the handle – the machine 'knew' what to do with the numbers stored on its wheels. You cranked a handle and read off the results. You did not need to understand the mathematical principle on which it was based or to have any knowledge of the internal mechanical workings in order to produce useful results. Each cycle of the engine produced the next result in the table with no intervention from the operator. The significance of the machine being automatic cannot be overstated. By cranking the handle, that is by exerting a *physical* force, you could for the first time achieve results that up to that point in history could only have been arrived at by *mental* effort – thinking. It was the first successful attempt to externalise a faculty of thought in an inanimate machine.

The Difference Engine demonstration piece was clearly not the first automatic machine of the industrial movement and it takes its place alongside clocks, trains, textile machinery and a host of other devices and systems. But with textile machines, trains and all the other wondrous contrivances that poured off the drawing boards of inventors and from the manufactories, the human activity they relieved or replaced was physical. Babbage's engine was a landmark in respect of the human activity it replaced. It was the first ingression of machinery into psychology. The first time he cranked the handle to take the machine through a calculating cycle marked the epoch of practical automatic computing. As the first automatic calculator it is the earliest surviving artefact in the field of machine intelligence.

From the start there has been a curious affinity between mathematics, mind and computing. Leibniz's work drew attention to the notion that mathematical thought might extend beyond numbers to principles of thought in general, and that mathematical logic might embody the rules of human reason. It is perhaps no accident that Pascal and Leibniz in the seventeenth century, Babbage and George Boole in the nineteenth, and Alan Turing and John von Neumann in the twentieth – seminal figures in the history of computing – were all, among their other accomplishments, mathematicians, possessing a natural affinity for symbol, representation, abstraction and logic. The relationship between the rules of logic and 'laws of thought' tantalised the thinkers of Babbage's generation. George Boole, a younger contemporary of Babbage, is perhaps best known to computer science students as the originator of Boolean algebra, a form of algebraic logic. The title of his major work, published in 1854, captures the interlacing of logic and mind which was a source of such speculation at the time – *An Investigation of the Laws of Thought*, *on which are founded the mathematical theories of logic and probabilities*. The work contains nothing we would now recognise as psychology. There is no theory of introspection, subjectivity or consciousness. It is a book of mathematical logic seen as a manifestation of regulated mental process.

Babbage's Engine gave new impetus to the notion of a 'thinking machine' and stimulated the debate about the relationship between mind and physical mechanism. The notion that the machine was in some sense 'thinking' was not lost on Babbage or his contemporaries: 'The marvellous pulp and fibre of a brain had been substituted by brass and iron, he [Babbage] had taught wheelwork *to think*, or at least to do the office of thought.' So wrote Harry Wilmot Buxton, a younger contemporary of Babbage and his posthumous biographer. Dionysius Lardner, who lectured and wrote at interminable length about the Engine, declared that 'the proposition . . . to throw the powers of thought into wheel-work could not fail to awaken the attention of the world'. In a sense, he was right, but just a century or so premature. The awakening came through electronics, a twentieth-century technology, rather than the cogs, wheels and levers of Babbage's time.

The small model Babbage made in the spring of 1822 has never been found, nor have any plans or drawings for it ever come to light, and it remains one of the unfound treasures of the history of the period. In the absence of a surviving relic or plans it is difficult to know whether this first machine was fully automatic. Certainly from the speed with which it reportedly rattled out results – 'producing figures at a rate of forty-four per minute' – it could not have been entirely manual.

The 'finished portion of the unfinished engine' delivered to Babbage in December 1832 is all that was ever assembled of the great Engine in Babbage's lifetime, and it remains the most substantial physical relic of one of the most remarkable engineering enterprises of the nineteenth century. It is the earliest surviving automatic calculator and one of the most celebrated icons in the history of computing.

The fuss we now make of the machine is informed by hindsight. Were it not for the prominence of the modern electronic computer and the way it has forced itself on our attention, it is unlikely that we would trouble so to excavate its meaning. But the celebrity of Babbage's Engine is not solely an artefact of modern history. In its day it was an object of wonder, and its demonstration excited philosophical and scientific speculation.

The successful construction of a piece of Difference Engine No. 1 was not the only rewarding event for Babbage in 1832. In that year the most substantial and successful of his six full-length works appeared – *On the Economy of Machinery and Manufactures*. This work, which has been described as a 'brilliant and utterly original foray into political economy', grew out of Babbage's systematic investigation of craft and manufacture prompted by the demands for precision in the construction of the Difference Engine. His knowledge of processes was encyclopaedic: from wire extrusion to the use of templates for vermicelli and ornamental pastry, from square glass bottles to umbrella handles.

Economy also considers the social organisation of labour, the size of factories, unionisation, overproduction and what we now refer to as 'technology transfer' to foreign countries. He advocates decimal currency, speculates about linking London and Liverpool by speaking-tubes, and foresees the exhaustion of coal reserves and the role of tidal power as a source of energy. He had boundless faith in the ingenuity of mankind, but commented wryly that if posterity failed to find a substitute for coal it 'deserved to be frostbitten'. The book was a tour de force and established Babbage as a major authority on the industrial movement. There is no trace here of the raging sarcasm and damaging diatribes that so marred *Decline*. The quality and seriousness of *Economy* went some way to rehabilitate his image in the scientific community after the damage done by his explosive demolition of the Royal Society two years earlier.

Part II

The Analytical Engine

Chapter 5

BREAKTHROUGH

The whole of arithmetic now appeared within the grasp of mechanism.

Charles Babbage, 1864

It took fully sixteen months to reach a settlement with the recalcitrant Clement. During this time Babbage was deprived of his drawings, and with work on the Engine at a standstill he had an enforced lay-off from the nuts and bolts of mechanical construction. Had Clement made a clean break sooner, or had the relationship continued undisturbed, Babbage would perhaps not have broken the pattern of preoccupation with the details of the machine's production. But with a long break from his earlier line of thinking, and once the drawings were back in his possession in July 1834, he began to question his original approach and returned to ideas that had lain dormant since the first conception of the machine. During this process Babbage conceived of a new machine which enthralled him for the rest of his life. Between summer 1834 and summer 1836, Babbage not only made the crucial breakthrough to a general-purpose computing engine but established most of the essential principles of its implementation.

The exact progression from the Difference Engine to the Analytical Engine is difficult to audit. Babbage kept a sort of laboratory journal which he called Scribbling Books or Sketch Books – scratch-pad notes, diagrams and jottings of his ideas. The bound volumes were for his private use, and the writing makes no concessions to the intrusive eyes of modern scholars eager to excavate his thoughts. Many of the crucial sheets are undated and the comments and diagrams are often cryptic or fragmentary, or presuppose a line of thought lost for all time. Maurice Wilkes, who in post-Second World War Cambridge pioneered the first usable electronic computer in Britain, the EDSAC, was among the first to evaluate Babbage's work in the modern era. Wilkes recalls his experience of Babbage's manuscript papers:

> So intimate is the impression created by Babbage's notebooks that one feels that one has strayed into his laboratory and, while waiting for him to come in, has started to read the papers that are lying about. They are not wholly intelligible, but one is sure that when he does come in he will tell one all about it.

With the single memorable exception of a visit to Turin in 1840 on which he addressed mathematicians and engineers, Babbage gave no public lectures on his new machine, nor did he address any scientific gathering to expound its wonders. He published little on his great invention other than a few general outlines, and these gave few technical details of its logic or mechanical execution. Certainly the number of his associates able to grasp the significance of the invention was few, and perhaps he despaired of being understood. Bruised as he was by the failure to complete the Difference Engine, and wounded by whispered ridicule and imputations of financial impropriety, it is possible that he was wary of exposing himself to further disappointment by advertising his work. He was in any case a rotten publicist, preferring to devote himself to the development of ideas rather than to their promotion.

With his published output so sparse and public explanations practically non-existent, there may arise the suspicion that Babbage's greatness is a backwards projection from our own age. The Science Museum Library in London holds a comprehensive manuscript archive of his work on the Engines from which it is clear that his unpublished work is as rich as his published work is meagre. His Scribbling Books run to between 6,000 and 7,000 sheets. There are some 500 large design drawings showing the mechanical details of the Engines' mechanisms, and about 1,000 sheets covered with his 'Notations' – symbolic descriptions of logical and mechanical flow using a sign language he invented. It is not a case of modern history seeing shapes where none exist to create vision out of suggestive hints, like finding prescience in the coded vagueness of Nostradamus. The scope and meticulous detail of this work is solid evidence of inspired accomplishment.

It was not until the late 1960s that this material was studied in any detail by scholars – over a hundred years after the ink had dried on the heavy parchment-like paper Babbage used. Nowhere does he describe the overall process of development or the progression of ideas that led him to the Analytical Engine designs. The evolution of his thinking has been pieced together as best it can from scraps, sketches, notebooks, and tantalising comments and diagrams. While the developmental line is tortuous, the outcome embodied in the hundreds of design drawings is unmistakable. It is only in the light of the studies since the late 1960s, viewed from the standpoint of modern electronic computing, that the startling extent of Babbage's achievement begins to emerge. What is simply astonishing is that the designs for the Analytical Engine embody in their mechanical and logical detail just about every major principle of the modern digital computer.

After the return of the drawings and parts from Clement, Babbage revisited his early ideas for the Difference Engine. One of the problems that had exercised him earlier was the need to reset the machine for each new run of calculations. At the start of a tabulation the initial values are entered into the Difference Engine by setting the figure wheels by hand. Once the Engine is set up in this way each turn of the handle produces the next result in the table, and the calculations follow in an unbroken sequence without further intervention by the operator. For some calculations the answers start to diverge from the true results as the table progresses – something like a delinquent shopping trolley staying on track for a while and then lurching off with increasing error. This is not a fault of the Engine. The calculation is often based on an approximation for the mathematical expression being tabulated, and the formula used is accurate only for a limited range of values; outside this range the results deviate from the true expression even though the results are correctly computed. As the calculation progresses a point is reached when the last digits in the result are no longer correct. Then, the longer the tabulation goes on the worse the deviations get. Even with the Engine functioning faultlessly, after a while the correctly computed results are not those desired.

It is possible to estimate in advance how many results in an unbroken sequence will be correct before the last digits fall into error. Once the run of calculations is outside the true range the Engine needs to be reset with a new set of starting values (called the pivotal values) for the next run. Keeping the calculation on track by frequent adjustment is inconvenient and also mathematically inelegant. Worse still, resetting the engine by hand is susceptible to precisely the sort of human errors the Engine was designed to eradicate.

During the earliest incarnation of the Difference Engine in 1822, Babbage noticed a particular property of a trigonometrical function (the sine function in this case) that would allow his engine to generate an endless sequence of correct values without having to halt it periodically to put the numbers back on track. By feeding values from one column to another during the course of the calculation, the machine would continuously adjust itself. The mathematics requires that the number fed back or forward be multiplied by a fixed constant. This was something he could not yet do, and the ideas were filed away at the time.

But with the 'beautiful fragment' of the Engine assembled by Clement in his hands, Babbage went back to these speculations and made some mechanical additions to the machine. He fitted some extra gear wheels and shafts that would allow single digits from one column to be fed automatically to the wheels of another column in an elementary exploration of the principle. (These additions are still visible on the surviving machine.) In the summer of 1834 he began to generalise a plan for feeding back multiples of values on one column to the wheels of columns elsewhere in the engine.

By a series of undocumented steps he arrived at a circular layout of columns that would allow the output to be fed continuously back to the input. He had closed the loop by folding the Engine back on itself, and he referred to this in a figurative flourish as 'the Engine eating its own tail'. He also refers to the self-generating properties of the Engine as 'a locomotive that lays down its own railway'. At this stage he had a dim sense of the grander possibilities. He later recalled:

> The circular arrangement of the axes of the Difference Engine round large central wheels led to the most extended prospects. The whole of arithmetic now appeared within the grasp of mechanism. A vague glimpse even of an Analytical Engine at length opened out, and I pursued with enthusiasm the shadowy vision.

The pursuit of the 'shadowy vision' unfolded between summer 1834 and 1836. In bursts of intense activity Babbage imparted to his new Engine a succession of features at which one can only marvel. He devised the first automatic mechanisms for direct multiplication and division. The process of division was particularly daunting, and he remarked that of the arithmetical operations of the machine 'none certainly offered more formidable difficulties'.

For division he designed the Engine to operate in an exploratory way. The machine performs an operation, examines the result, and takes an appropriate action depending on the outcome. Specifically, the mechanism for division performs a series of tentative subtractions until the result goes negative. The detection of the minus sign indicates that there has been one subtraction too many, and the next step is to restore the correct number by an addition. The result is then multiplied by ten and the process of repeated subtraction is repeated. By keeping count of how many subtractions are executed at each stage before the sign change, the digits of the result of the division are generated in turn.

The control mechanism of the Engine executes these operations automatically. The sequence of smaller operations required to effect an arithmetical operation was controlled by massive drums called barrels. The barrels had studs fixed to their outer surface in much the same way as the pins of a music box drum or a barrel organ. The barrels orchestrated the internal motions of the engine. As a barrel turned, the studs activated specific motions of the mechanism and the position and arrangement of the studs determined the action and relative timing of each motion. The act of turning the drum thus automatically executed a sequence of motions to carry out the desired higher-level operation. The process is internal to the Engine and logically invisible to the user. The technique is what in computing is now called a 'microprogram' (though Babbage never used the term), which ensures that the lower-level operations required to perform a function are executed automatically.

From his survey of workshops and factories and his practical experiences in his own workshop, Babbage was a reasonably accomplished amateur machinist. In the early 1820s he had his own multi-purpose lathe made to order by Clement. But he had learned to his cost to husband his time and mental energy: 'I at length laid it down as a principle – that, except in rare cases, I would never do anything myself if I could afford to hire another person who could do it for me.' True to his maxim, in the autumn of 1834 he hired Charles Godfrey Jarvis at his own expense as his principal draughtsman. It was Jarvis who had worked on the Difference Engine in Clement's firm and had warned Babbage of his suspicions that Clement had a vested self-interest in delaying completion of the Difference Engine. Babbage also hired two or three assistants. He was competing for skills with the booming growth of the railways: skilled designer-draughtsmen of Jarvis's calibre were in demand, and at one stage Jarvis was tempted by a lucrative offer abroad. Babbage reckoned that he would need to up his rate to about a guinea a day to beat off the competition. He consulted his mother who, as Babbage relates, counselled him thus:

> My dear son, you have advanced so far in the accomplishment of a great object, which is worthy of your ambition. You are capable of completing it. My advice is – pursue it, even if it should oblige you to live on bread and cheese.

Babbage was well able to afford Jarvis's rates without compromising the quality of his dining table. But he was clearly gratified by this ringing endorsement from his mother, which so contrasted with the dismissiveness of his late father and Jarvis became the beneficiary of the handsome new rate. The arrangement to retain him on a more secure basis was consolidated in 1835. Many of the fine drawings of the Analytical Engine are by Jarvis's hand.

With mechanisms for multiplication and division under his belt, Babbage had in fact designed a general-purpose four-function calculator capable of the basic arithmetical operations – addition and subtraction (which he had already cracked for the Difference Engine), multiplication and division. He became obsessed with the need for speed, and agonised endlessly over optimising the mechanisms to minimise the execution time of the various operations. 'The whole history of the invention has been a struggle against time,' he wrote in 1837. Reducing execution time was a ruling principle of the design, and he endlessly refined and then discarded techniques if they could be improved.

> As soon as any contrivance has been made . . . the question has always arisen: Can it not be executed in

> less time by some other contrivance? Thus every advance has but raised up a new object of rivalry, itself to be superseded by some more rapid means, nor can I hope that I have nearly reached the limit. If I have approached it, it is more than I have a right to expect as the pioneer in this difficult career. Another age must be the judge of that as well as of the other questions relating to the engine . . . The reader . . . will doubtless be surprised . . . at the lavish rejection of inventions which has taken place in order to achieve rapidity in computation.

Multiplication and division were irreducibly time consuming since the process involved a sequence of repeated operations. To free up time for the longer operations he set about reducing the time of arithmetical addition. The technique used in the Difference Engine, though ingenious and reliable, was too time consuming. We have seen that the desktop mechanical calculators of the time were limited in how many digits they could handle because of the difficulty of carrying tens. We have also seen that Babbage cracked the problem in Difference Engine No. 1, which featured numbers with sixteen and later with twenty digits. The mechanism that allowed him to so extend the precision of his machine was not only highly effective, but bestows one of the most arresting visual features on the motions of the mechanism. Say, for example, that we need to add two numbers, each of which has thirty digits. Each digit of each number has its own figure wheel, and these are stacked vertically in columns with the units below, tens next, and so on. Each column has thirty wheels. What Babbage did was to allow the engine to execute the addition of all the digits at once by meshing the two columns digit for digit. If the numbers on two meshed figure wheels add up to more than ten, then during the addition one of the wheels will advance from '9' to '0'. But he did not use this overrun to directly drive the next higher wheel on by '1' – he captured the motion instead. As the wheel passed from '9' to '0', a peg on the wheel nudged a lever which latched into a new position. Each wheel was fitted with a latching device to warn which digit positions still needed to carry tens to the next wheel up. If a latch was set, Babbage referred to it as being in its 'warned' state. If the addition of two digits was less than ten, then the latch was left unchanged in its 'unwarned' state.

The next phase of the cycle polls each of the latching devices in turn by an arm sweeping past each latch. If the latch is warned, the trajectory of the arm intersects with the stationary lever and the act of sweeping the latch lever aside advances by one position the next wheel in the stack to carry the tens. If the latch is unwarned, the arm and the lever miss each other and no action is carried out. The polling of each of the latches is staggered a little in time so that each interrogation occurs slightly after the last. Babbage called this method of carrying tens 'successive carry' since the method deals with each digit position in turn.

The arms that poll and sweep are in a vertical stack on a shaft. Each arm is set in a position slightly lagging behind the one above it so that together they form a helix, rather like a circular staircase. In Difference Engine No. 1 the density of parts is such that it is almost impossible to see this sweeping motion. But in his later engine, Difference Engine No. 2, the motions are exposed, and the rotation of the helices during the calculation produces a spectacle of arresting beauty.

Just as electronic engineers over a century later would talk about latches acting as one-digit memories, Babbage writes of the latching action in his carry mechanism being akin to an act of memory, and the polling action to that of recollection:

> Now there is in this mechanism a certain analogy with the act of memory. The lever thrust aside by the passage of the tens, is the equivalent of the note of an event made in the memory, whilst the spiral arm, acting at an after time upon the lever put aside, in some measure resembles the endeavours made to recollect a fact.

The great advantage of the technique of successive carry is that there is never any more force exerted on the shaft that sweeps the latches than it takes to move one figure wheel one position on. By separating the addition part of the cycle and postponing the carriage of tens, Babbage was able to store and operate on numbers twenty, thirty and fifty digits long – five to ten times longer than any device devised before. But there was a price. Time. Since each digit position is polled in turn the whole operation is costly in execution time, and the more digits in a number, the longer the process takes. The technique, elegant and ingenious as it is, was fast enough for the Difference Engine but not fast enough for the Analytical Engine.

Babbage went to extraordinary lengths to reduce the time taken for the carriage of tens, fighting for time to accommodate the operations of multiplication and division which themselves could be reduced no further. To hasten the process of carrying tens he designed a mechanism that anticipated when a carriage would be needed and executed all the carries in one operation. It was an astonishing feat of intellect as well as mechanical ingenuity. He was so absorbed in the exploration and realisation of the technique of the anticipating carriage that one of his workmen feared for his master's sanity. Thirty years after the event Babbage recalled:

> The most important part of the Analytical Engine was undoubtedly the mechanical method of carrying the tens. On this I laboured incessantly, each succeeding improvement advancing me a step or two. The difficulty did not consist so much in the more or less complexity of the contrivance as in the reduction of the *time* required to effect the carriage. Twenty or thirty different plans and modifications had been drawn. At last I came to the conclusion that I had exhausted the principle of successive carriage. I concluded also that nothing but teaching the Engine to foresee and then to act upon that foresight could ever lead me to the object I desired, namely, to make the whole of any unlimited number of carriages in one unit of time. One morning, after I had spent many hours in the drawing-office in endeavouring to improve the system of successive carriages, I mentioned these views to my chief assistant, and added that I should retire to my library, and endeavour to work out the new principle. He gently expressed a doubt whether the plan was *possible*, to which I replied that . . . I should follow out a slight glimmering of light which I thought I perceived.
>
> After about three hours' examination, I returned to the drawing-office with much more definite ideas upon the subject. I had discovered a principle that proved the possibility, and I had contrived a mechanism which, I thought, would accomplish my object.
>
> I now commenced the explanation of my views, which I soon found were but little understood by my

> assistant; nor was this surprising, since in the course of my own attempt at explanation, I found several defects in my plan, and was also led by his questions to perceive others. All these I removed one after another, and ultimately terminated at a late hour my morning's work with the conviction that *anticipating* carriage was not only within my power, but that I had devised one mechanism at least by which it might be accomplished.
>
> Many years after, my assistant, on his return from a long residence abroad, called upon me, and we talked over the progress of the Analytical Engine. I referred back to the day on which I had made that most important step, and asked him if he recollected it. His reply was that he perfectly remembered the circumstance; for that on retiring to my library, he seriously thought that my intellect was beginning to become deranged.

Babbage worked through that day, ending an eleven-hour stint at seven in the evening, at which point he broke off to keep a dinner appointment with friends in Park Lane. Intoxicated with his breakthrough, he remarked to his host that the rush of invention 'had produced an exhilaration of the spirits which not even his excellent champagne could rival'. But this inventive high was cruelly crushed by yet another personal tragedy. It was around that time in 1834 that Georgiana, his only daughter, died while still in her teens. Babbage had doted on her, and it is quite possible that he threw himself into his work so energetically in order to numb himself, as he had done after the death of his wife in 1827. If ever he confided about his grief in letters, none seem to have survived.

Babbage writes of 'teaching the Engine to foresee'. Elsewhere he talks of the Engine 'knowing'. He was clearly sensitive to the appropriateness of using language in this way, and evidently felt that anthropomorphising mechanism required some justification or excuse:

> The analogy between these acts and the operations of mind almost forced upon me the figurative employment of the same terms. They were found at once convenient and expressive and I prefer continuing their use rather than substituting lengthened circumlocutions.

The anticipating carriage mechanism reduced the length of time taken to effect the carriage of tens. There is no direct modern counterpart to the anticipating carriage mechanism, and despite its name it is not what we would now call a 'look-ahead' technique. Unlike the successive carry method, in which the latches are serviced one at a time, the anticipating carriage mechanism services all carries in one motion. The time taken for the carriage of tens is therefore independent of the number of digits in the column – an elegant improvement on the technique of successive carry, in which the larger the number of digits, the taller the column of wheels and the longer it takes to poll them. The new design also handles secondary carries (carries that result from carries).

The design of the anticipating carriage was a brilliant technical coup, and Babbage was delighted. There were other more far-reaching features of the Analytical Engine that he rated less highly simply because they had come to him more easily. Babbage's success in optimising the carriage of tens was not only a technical breakthrough but acted as a psychological boost to his confidence in his ability to accomplish ever more ambitious powers for the Engine. He wrote that after the episode he 'felt renewed power and increased energy to pursue the far higher object I had in view'. It is this 'higher object' on which Babbage's fame and reputation as the first computer pioneer largely rests.

The anticipating carriage mechanism prompted a major step in the evolution of the Analytical Engine. Babbage found that ten times as much machinery was required for the carriage of tens as was required to execute the basic operation of addition. Instead of peppering the machine with expensive carriage mechanisms wherever they were needed, he saw that there were significant economies to be had by centralising expensive and complex machinery in one place and transferring information to the central apparatus for processing. So he split the machine into two parts. The section of the Engine which performed the various arithmetical operations he called the Mill, and the area where the numbers were kept he called the Store – terms borrowed from the textile industry. The Store consisted of columns of figure wheels which held the numbers to be operated on. It corresponds to what we would now call memory. Numbers in the Store are transferred to the Mill by long toothed racks for processing, and the results returned to the Store on completion. The Mill is directly analogous to the modern notion of an arithmetic unit or central processor.

It is a startling fact that the logical and physical separation of the Store and Mill (memory and central processor) is a fundamental feature of the modern electronic digital computer. The logical blueprint for the modern electronic computer was first publicly articulated by John Von Neumann, a Hungarian-born mathematician. In a seminal paper written in 1945, Von Neumann describes the internal organisation of the electronic computer. This layout, which became known as 'Von Neumann architecture', has dominated computer design to the present day, and is incorporated in just about every computer around. A feature of this architecture is the separation of the central processor from the memory – a feature explicitly used by Babbage over a century earlier.

With the major conception of the Engine and its design shaping up fairly well, Babbage turned his attention to ways in which the machine would output its results. He proposed a full range of automatic output devices, including a printer, a card-punch and curve-drawing apparatus, all directly coupled to the machine. If a few copies of the results were needed, the printing apparatus would press numbers through sheets interleaved with carbon paper. If multiple copies were needed, the apparatus would impress results on copper plate or papier mâché to produce a mould from which conventional stereotype printing plates could be made. Output data could also be punched out directly onto number cards coded so that the pattern of holes represented the value of the number. The number cards could be used to store previously computed results or to dump intermediate results that would be used in a later calculation. The cards would then be read by a card reader, one of the input devices. With the ability to output results on punched cards, Babbage realised that the printing apparatus need no longer be integral with the calculating section, and that printing and stereotyping could be performed off-line in a different location and at a different time.

So far Babbage had designed a machine with a repertoire of the four basic arithmetic functions. The internal architecture was established, and the cycle time of the machine had been reduced as far as he was able. He also had clear ideas about how to produce output in various forms. The problem now was how to instruct the machine to perform its various operations. Changing the studs on the barrels was an ungainly and impractical prospect. If there were separate barrels for each operation he would need a way of instructing the machine as to which operations he wished to perform and in what order. In short, Babbage needed a method of *programming* the engine (though at no stage did he use the word to describe the process).

In June 1836 Babbage opted for punched cards to control the machine. The principle was openly borrowed from the Jacquard loom, which used a string of punched cards to automatically control the pattern of a weave. In the loom, rods were linked to wire hooks, each of which could lift one of the longitudinal threads strung between the frame. The rods were gathered in a rectangular bundle, and the cards were pressed one at a time against the rod ends. If a hole coincided with a rod, then the rod passed through the card and no action was taken. If no hole was present then the card pressed back the rod to activate a hook which lifted the associated thread, allowing the shuttle which carried the cross-thread to pass underneath. The cards were strung together with wire, ribbon or tape hinges, and fan-folded into large stacks to form long sequences. The looms were often massive and the loom operator sat inside the frame, sequencing through the cards one at a time by means of a foot pedal or hand lever. The arrangement of holes on the cards determined the pattern of the weave.

As well as patterned textiles for ordinary use, the technique was used to produce elaborate and complex images as exhibition pieces. One well-known piece was a shaded portrait of Jacquard seated at table with a small model of his loom. The portrait was woven in fine silk by a firm in Lyon using a Jacquard punched-card loom. The image took 24,000 cards to produce, and each card had over 1,000 hole positions. Babbage was much taken with the portrait, which is so fine that it is difficult to tell with the naked eye that it is woven rather than engraved. He hung his own copy of the prized portrait in his drawing room and used it to explain his use of the punched cards in his Engine.

The delicate shading, crafted shadows and fine resolution of the Jacquard portrait challenged existing notions that machines were incapable of subtlety. Gradations of shading were surely a matter or artistic taste rather than the province of machinery, and the portrait blurred the clear lines between industrial production and the arts. Just as the completed section of the Difference Engine played its role in reconciling science and religion through Babbage's theory of miracles, the portrait played its part in inviting acceptance for the products of industry in a culture in which aesthetics was regarded as the rightful domain of manual craft and art.

Babbage and his Engines were celebrated attractions and sometimes featured in the VIP London tour. In 1842 the Queen's uncle, Count Mensdorf, asked to see Babbage's Engine, and an appointment was made for Mensdorf and the Duke of Wellington to visit Babbage at his house. The day before the visit Babbage received a note informing him that Prince Albert, a committed champion of the industrial arts, was now joining the group and that the time of the arrival was to be an hour later than agreed.

The late change in the arrangement irritated Babbage, but he was determined to do his duty as the host to royalty to whom he does not appear to have been too well disposed. On the arrival of his guests, Babbage drew their attention to the Jacquard portrait. 'Oh! that engraving?' remarked Wellington. 'No,' said the Prince, 'it is not an engraving. I have seen it before.' Babbage was completely won over. He wrote that he 'felt at once that the Prince was a "good man and true"', and devoted himself to satisfying the Prince's interest and curiosity about the Engine. Babbage was a man of extremes. His friends could do no wrong and his enemies could do no right.

The Prince's remark was all it took to secure a lasting allegiance that rank alone was unable to command. Babbage's boyish loyalties have an embarrassing charm.

With Jacquard's loom as the model, Babbage used pasteboard cards to instruct the Engine and designed a card reader that would step through the sequence of cards and read the hole pattern. As in the Jacquard apparatus, the cards were pressed against a matrix of rods. Where there was no hole the rod was activated to initiate an operation. Babbage proposed several varieties of card. Operation Cards told the Mill which operation to perform – addition, subtraction, multiplication or division. A string of Operation Cards constituted a set of instructions to be executed in a fixed sequence and which for all practical purposes could be extended without limit. There were Variable Cards which specified from where in the Store the number to be operated on was to be fetched and where in the Store a result was to be returned. A third type of card was the Number Card, which held numerical data and had a variety of uses. Instead of setting the initial values in the Store by turning wheels by hand, this data could be input from cards into the store automatically. The Number Cards could also act as a reserve memory whenever the Store did not have enough room: the overflow data could be punched onto cards and then retrieved as and when needed.

Punched cards had the additional advantage of being a permanent rather than a temporary record. When a number is read from a figure wheel by rotating it to zero, the original number is lost. Unless a separate operation is undertaken to restore the number, such read-out is destructive. An advantage of coding information on punched cards is that the hole pattern is permanent and allows non-destructive read-out. Number Cards could also be used to store the values of repeatedly used physical constants or pre-computed logarithms, for example, which could be kept as a library to be consulted by the Engine as the need arose. Once such values are correct there would be no risk of corruption through human error. Babbage also saw deeper logical implications of these cards. Since Number Cards could act as an extension of the Store there was no limit to the amount of data accessible to the machine, and a complex calculation could freely exchange information to and from a limitless supply of cards.

Another feature developed in this first ground-breaking period added a final finesse – the capacity of the Engine to automatically repeat a sequence of operations a predetermined number of times. To accomplish this, Babbage introduced a fourth class of card called the Combinatorial Card. The ability to loop back and iterate a set of operations has great importance for calculations involving successive approximations where the result converges progressively with each repetition of the calculation.

The machine could now execute what today we would call programs – sequences of instructions that used the internal repertoire of operations in any desired order. There was no attempt to store the users' programs internally. There were internal 'microprograms' controlled by barrels, and data was held in the columns of figure wheels that made up the Store. But the user programs were stored on punched cards external to the machine. There is no evidence of the 'stored program' concept which is a hallmark of the modern digital computer, though this should in no way diminish the monumental accomplishment represented by the new Engine.

Babbage reckoned that the needs of science would be met for a long while to come by numbers with no more than thirty digits, though he did not constrain the machine to these limits. Parts of the machine cater for forty and even fifty digits, and in some situations he made allowance for double-precision arithmetic using hundred-digit numbers. Given that the Engine was a lumbering mechanical beast, it was surprisingly fast. Babbage estimated that the Analytical Engine could in one minute divide a hundred-digit number by a fifty-digit number, or multiply two fifty-digit numbers, or perform sixty additions or subtractions as well as print the results.

The Difference Engine and Analytical Engine are decimal digital machines. They are decimal in the sense that they use our familiar ten numbers '0' to '9', and they are digital in the sense that only whole-number digit values are recognised as valid. For example, the number '2.5' is represented in the engine by a figure wheel rotated to show '5' and the wheel immediately above showing '2'. Each digit has its own wheel. A wheel halfway between digits is invalid, and the control system of the engine is designed to interpret this as an indeterminate value. So if you attempted to represent the value '2.5' by setting the units wheel halfway between '2' and '3', the engine would jam.

Since the whole rationale for the Engines was the absolute integrity of results, Babbage went to extraordinary lengths to ensure that the machine was immune from deliberate or inadvertent derangement. He incorporated a system of locking devices that immobilise the figure wheels during periods of the cycle in which they are inactive. One technique uses wedges which are driven automatically between the teeth of wheels to lock them during dormant parts of the cycle. The wedges are withdrawn to release the wheels only for the short window in the cycle in which the wheels might be moving. If a wheel does stray into an indeterminate position, when the wedge tries to re-enter it is met end-on by a tooth. The parts clash and the engine jams. Jamming is part of error-detection, and while frustrating in practice it is a reliable indication of malfunction. The insertion of the wedges also has a self-correcting effect: any small departure from fixed one-digit increments is automatically removed by the restoring action of the wedge. The electronic equivalent of the action of the wedge is what electronic engineers would now call 'pulse-shaping'. Babbage adhered rigorously to the principle that 'every motion shall be of such a kind that the Engine shall either break itself or stop itself or execute the intended motion'.

He went beyond immobilising inactive wheels by designing mechanisms to ensure that a figure wheel can be advanced only by input from a legitimate source: from a wheel alongside during addition, for example, or from a wheel below as a result of carrying tens. Such was his confidence in the safeguards against corruption he went so far as to challenge an operator to deliberately falsify the numbers on the wheels by arbitrarily turning them even while the engine was busy calculating, and he defied anyone to compromise the production of the correct result by such intentional mischief. Some of the protective mechanisms are of extraordinary subtlety, and their function was not fully appreciated until a full machine finally came to be built.

Although the Difference and Analytical Engines are decimal machines, Babbage did not take the decimal number system for granted. He considered using a binary number system as well as other number bases, including 3, 4, 5, 12, 16 and 100. His choice of the decimal system was based on engineering efficiency. The higher the base, the larger the number of entities that need to be distinguished, and physical discrimination becomes increasingly difficult. Lower number bases make discrimination between values easier (the binary system requires discrimination between only two states), but lower bases demand more mechanical parts to represent a given number and, as we have seen, the difficulty of manufacturing near-identical parts was a weakness of the technology of the time. So he opted for the conventional base of '10' out of engineering convenience and mechanical economy, not out of any assumption about the sanctity of tens.

The Difference Engine crunched numbers the only way it knew how: it performed repeated addition according to the method of finite differences. Any numbers entered into the machine are treated in exactly the same way. The Difference Engine is incapable of performing any operation other than that dictated by its wheelwork. For all the ingenuity of its execution and the historic fact of its being automatic, it is what we would now call a calculator rather than a computer: it is capable only of the fixed, specific and predetermined functions allowed by its construction. What we now take to be a computer is necessarily a general-purpose machine capable of being programmed by the user. And the Analytical Engine pretty well fits the bill.

Chapter 6

APPLAUSE IN TURIN

It will not slice a pineapple.

Charles Babbage, 1852

The conception and design of the Analytical Engine ranks as one of the startling intellectual achievements of the nineteenth century. Maurice Wilkes describes Babbage's work on the Analytical Engine as possessing 'vision verging on genius'.

The features Babbage invested in the Engine are astonishing. It could be programmed by the use of punched cards. It had a separate 'memory' and 'central processor'. It was capable of 'looping' or 'iteration' (the ability to repeat a sequence of operations a programmable number of times) as well as conditional branching (the ability to take one course of action or another, depending on the outcome of a calculation – IF . . . THEN statements). It incorporated 'microprogramming' as well as 'pipelining' (the preparation of a result in advance of its need) and catered for the use of multiple processors to speed computation by splitting the task – the basis of modern parallel computing. It also featured a range of input and output devices, including graph plotters, printers, and card readers and punches. In short, Babbage had designed what we would now call a general-purpose digital computing engine.

It was not just the logical conception of the Analytical Engine that was unprecedented, its physical scale was simply stupendous. The central section of the Mill would be some fifteen feet tall and six feet in diameter. The length of the Store depended on its capacity. A modest configuration with a capacity for storing 100 fifty-digit numbers would extend the machine to about twenty feet in length, making it the size of a steam locomotive. Babbage thought big. He wrote of machines with a storage capacity of 1,000 numbers. A Store of that size would extend the machine to over one hundred feet in length, and there seems little prospect of such a mammoth mechanism ever being driven by hand. Had it been built, 'calculating by steam' may well have been a prophetic prediction rather than a figure of speech.

The most original period of development on the Analytical Engine ended in the summer of 1836. After that Babbage elaborated, refined and revised the design continuously, but the major features were already largely in place.

It is customary to refer to the Analytical Engine as though it were a physical thing. This is a linguistic shorthand. It was never built, and remains an abstraction captured on the drawing board and in manuscripts. It is also customary to refer to the Analytical Engine in the singular, as though the drawings represent a single convergent design. This too is a convenience of language. It is more correct to speak of 'Analytical Engines' in the plural, as there were several specialised versions optimised for particular applications, and designs were constantly being modified and improved. Before 1840 there was the Great Analytical Engine. Later there was a sawn-off, simplified and slower Small Analytical Engine as well as various transitional variants. Although there are discernible milestones, there is no single design that can be held aloft as a specification of the definitive machine.

Babbage worked in almost complete isolation. Wilkes, who was among the first to study Babbage's unpublished work, makes a bleak observation:

> Ever since going through Babbage's notebooks, I have been haunted by the thought of the loneliness of his intellectual life during the period when, as he later tells us, he was working 10 or 11 hours a day on the Analytical Engine.

During the years of development Babbage certainly discussed practical aspects of his schemes with his workmen, whom he hired at his own expense and supervised in his private workshop. But there is no evidence that there was anyone with whom he shared as an equal the intellectual content of his work. He was ordinarily a prolific correspondent, and it is curious that there are practically no surviving letters to friends or colleagues about his new pursuit. In the early 1840s, on Babbage's insistence, his sons Henry and Dugald, then in their late teens, spent time – two or three times a week – in the drawing office and workshop in Devonshire Street learning workshop and drafting skills from their father's assistants. Henry later acquired a sound grasp of aspects of the Difference and Analytical Engine designs, and came to form a close bond with his father during furloughs from military service in India. But during 1834, the *annus mirabilis* of his father's invention, Henry was a lad of ten and too young to be a technical confidant. Several years after the pioneering foundations were laid,

Babbage explained many of his ideas to Ada Augusta Lovelace, the only legitimate daughter of the poet Byron, who befriended Babbage and took an avid interest in his work. But for the most part the conception and creative development of the Analytical Engine took place in a private and personal realm, largely unshared.

While Babbage courted celebrity and recognition for his Difference Engine, when it came to the Analytical Engine he was more reticent in his prosecution of public or indeed scientific acknowledgement. The original stimulus for the Difference Engine was to eliminate the risk of human error in the production of printed mathematical tables. In so far as Babbage and his supporters cared to justify the need for mechanical calculating engines, their advocacy was clothed in arguments about the benefits to navigation. It is difficult to know to what extent the argument of need served as the continuing motive behind Babbage's unrelenting work on the Difference Engine, or whether need served only as a launch pad for work driven by his relish of mechanics and logic. In the case of the Difference Engine he was content to let others, Lardner for example, trumpet the horrors of tabular errors. But when it came to the Analytical Engine the issue of practical utility barely features. He was entirely seduced by the intellectual quest and propelled by an unremitting fascination with its mechanical realisation. His great invention was not a response to any identified need. The rhetoric of the time makes reference to the grand enterprise of science and the noble extension of the feeble capacities of the human mind by machine, but these are postures, not motives. Babbage was driven by the exploration of the possible. He had glimpsed some profound vision, and he beckons to us over the heads of his contemporaries.

But he was not entirely reclusive about his new machine. He wrote to his friend Adolphe Quételet, a Belgian statistician.

The letter, read to the Royal Academy of Sciences in Brussels in May 1835, was the first written leakage about the new machine to the world of science. As Babbage informed his friend:

> During the last six months I have been contriving another engine of far greater power. I have given up all other pursuits and am making drawings of it and advance rapidly but it is most improbable that it will be executed here. I am myself astonished at the powers I have given it. A year ago I would not have thought this possible.

He goes on to say that he had overcome the greatest difficulties of the invention and that a few more months would be needed to complete the detailed design. He clearly believed that he had cracked the worst of the problems and that an end was in sight. In a letter to an American friend in August 1835 he acknowledges the contribution of his guinea-a-day designer-draughtsman, Jarvis, and the convenience of not having to travel to Clement's premises:

> I have had a draftsman constantly at work since September and I have given up all other pursuits for the sake of this, and the progress I have already made surpasses my expectation, it is indeed more than I ever before made in several years. This arises partly from a more enlarged experience, partly by having a better tempered draftsman, partly by having all my drawings in my own house instead of at a distance of four miles. My intention is to finish a complete set of mechanical drawings of the new engine and a variety of Mechanical Notations to explain its operation.

The Mechanical Notation was a novel design aid devised by Babbage to help visualise the operation of his elaborate mechanisms. The unprecedented complexity of the engines made new demands on design technique and on existing modes of representation and record. He used standard drafting conventions to depict parts and assemblies. These consist of views of mechanisms drawn as seen from above (plan views) and from the front and side (elevations). The depiction of machinery in such drawings is necessarily static. To explore the trajectory of motions requires laboriously repeating views with moving parts in different positions – much like the hand-drafting of early animated cartoons by Walt Disney. To help with dynamic behaviour he habitually made small experimental pieces to explore and verify the operation of particular mechanisms. But the whole machine with all its subsystems was so intricate, and the chain of events so elaborate, that keeping track 'baffled the most tenacious memory'. To assist in marshalling the mechanism and the motions of interrelated parts he developed an elaborate system of signs and symbols which he called the Mechanical Notation.

The need for some new design aid arose early on in the work on Difference Engine No. 1, and he published a paper in the *Philosophical Transactions of the Royal Society* in 1826 in which he accounts for the genesis of the scheme:

> The difficulty of retaining in the mind all the cotemporaneous [*sic*] and successive movements of a complicated machine, and the still greater difficulty of properly timing movements which had already been provided for, induced me to seek for some method by which I might at a glance of the eye select any particular part, and find at any given time its state of motion or rest, its relation to the motions of any other

> part of the machine, and if necessary trace back the sources of its movement through all its successive stages to the original moving power. I soon felt that the forms of ordinary language were far too diffuse to admit of any expectation of removing the difficulty, and being convinced from experience of the vast power which analysis derives from the great condensation of meaning in the language it employs, I was not long in deciding that the most favourable path to pursue was to have recourse to the language of signs.

The Mechanical Notation is not a calculus in the sense that it does not allow the calculation of motion or its effects. It is a descriptive system that precisely records the way parts are intended to interact. One form of the Notation consists of timing diagrams which show how different motions are phased and harmonised. Another form resembles what we would now liken to logical flow diagrams. Yet a third form, extensively used to annotate the design drawings, uses a letter of the alphabet to identify a part. If a part is a piece of the fixed frame, for example, it is depicted by an upright capital F. If the part is movable, its letter label is italicised, say *T*. At least three alphabets were used. Numerous subscripts and superscripts dotted around the letter indicate the type of motion of the part it identifies: whether circular or linear, continuous or intermittent, how many other parts connect with it and which other parts drive it or are driven by it. The system of signs extends to show whether the motion of one part always entails the motion of another, as it would if they were permanently linked, or whether there is partial entailment as, say, in the case of a slotted hole. In its variations and versatility, the system is thoroughly baroque.

Babbage was inordinately proud of the Mechanical

Notation and regarded it as among his finest inventions. He used it to optimise designs, minimise the number of parts, detect redundancy and manipulate long chains of events in an abstract shorthand all his own. He did not see its application confined to machinery, but viewed it as a universal abstract language of interaction applicable equally to the circulation of the blood, respiration in animals, the organisation of factories, and combat by sea or land.

With such grand ambitions for his prized creation, he was hurt when the Mechanical Notation failed to win the first Royal Medal awarded by the Royal Society in 1826. One lucky recipient of that year's medal award was John Dalton, for his work on atomic theory. Babbage was outraged, and he publicly denounced the Royal Society for violating its own rules by giving an award for work done twenty years previously. His protest was not through any lack of generosity towards Dalton; the snub had touched the raw nerve of lack of recognition, and the grievance festered for years. But his faith in the Notation was unshakeable. In his seventies he wrote:

> I look upon it [the Mechanical Notation] as one of the most important additions I have made to human knowledge. It has placed the construction of machinery in the rank of a demonstrative science. The day will arrive when no school of mechanical drawing will be thought complete without teaching it.

For all his hopes and for all the undoubted use it served in the evolution of the Engines, the Notation was widely ignored by his contemporaries and, with the exception of a few hardy souls (notably his son Henry Prevost in the nineteenth century and Allan Bromley, the Australian computer scientist and Babbage engine scholar, in the twentieth), the detail of its content has enjoyed spectacular obscurity ever since. To us the Mechanical Notation appears as an oddity of a formidably individual mind.

Babbage had hopelessly underestimated the likely completion time for the Difference Engine, which remained unfinished after over a decade of work. He was nothing if not consistent, and continued to make equally hopeful predictions about progress on the new Engine. In a letter of August 1835 to an American friend, he writes that 'I have already got over all the difficulties of the first order and most of those of the second and feel very confident that in twelve months more, if I can carry it on as I have done, that I shall have completed the drawings so that the invention can not ever be lost.' Babbage spent nearly thirty years of his remaining forty-five developing and refining the designs. He never did complete the task.

Despite his optimism, Babbage made no concerted attempt to build the new machine. He did not lobby government for funds, and he regarded the project as too costly to undertake himself. He also did not see how the Engine could provide a financial return for any potential investor. His pursuit of practical detail came not from any clear ambition to build the machine, but rather from his drive for the mastery of technique and the relish of the intellectual exploration of an extraordinary new world in which he was the first inhabitant. He worked *as though* the machine might one day be built, and some of the specifications are sufficiently detailed to enable construction to proceed, but there was no clear intention to pursue any practical programme of realisation.

In the meantime, the Difference Engine was ignored. In the same year there was a Parliamentary debate on the Civil Contingencies fund which bankrolled the extravagant expense of the project. Several ventures, Babbage's included, were targeted for criticism as 'unprincipled waste and squandering public money'. The great Difference Engine remained unfinished business and a painful reminder of unfulfilled hope as well as public reproof and unfounded suspicion of personal profit.

The fate of the Difference Engine irked Babbage during the years he worked on the new machine. Every now and again he rattled the skeleton in the cupboard by writing to the Government in sporadic attempts to clarify the situation and establish the Treasury's perception of his obligations to the abandoned venture. In December 1834, a few months after his earliest inklings of the new vistas offered by the Analytical Engine, he wrote to the Duke of Wellington, then Foreign Secretary in the new Tory Government. The letter is grumpy, aggrieved and borders on belligerence. Babbage complains of personal and professional sacrifices, official neglect, false allegations of financial self-interest, and expresses his indignation at the lack of acknowledgement. The letter is also self-pitying. 'If it has consumed thirteen years of my life instead of two or three, it will be said that I must suffer the inconvenience . . . I was ignorant of that which no human being could foresee; I have suffered the consequences.'

He also commits what could be seen as an honourable but fatal strategic blunder. He tells the Duke that he is working on a new machine. Babbage does not describe it as revolutionary or different in kind from the Difference Engine but as an extension of it. He tells the Duke that he feels honour bound to inform the Government of the new development as concealment would be morally wrong. But he makes no specific request for support, and no recommendation for further action. It is not at all clear what the letter was intended to achieve.

His reference to the new direction of the work is self-contradictory. He refers to 'a totally new engine possessing much more extensive powers', but also states that it is not intended to supersede the earlier engine. He also notes that none of the devices and contrivances of the new machine were used in the old one. Was he suggesting that Difference Engine No. 1 be abandoned and that he start afresh? Was he asking in a coded way for renewal of financial support for the new Engine? And he shows no awareness that, to those who supported him, the sorry saga of the Difference Engine was an expensive embarrassment. In practical terms the mountain of money had begat very little, and Babbage seemed insensitive to the problem of his own credibility.

The emotional confusion in Babbage's letter and its mixed message were not helped by the new round of musical chairs being played by Whigs and Tories at Westminster. There were four changes of government between July 1834 and April 1835, and Babbage waited for more than a year for a reply to his long letter to Wellington. When a reply did come it was clear that his submission to Wellington had indeed been taken as a request for support – something Babbage hotly denied. By this time he had a clearer idea of the powers of the Analytical Engine, and in his reply to the Chancellor dated 2 February 1836 he compounded the confusion by reversing his earlier assertion that the new Engine would not supersede the old:

> [it] performs all those calculations which were peculiar to the old Engine both in less time and to a greater extent – in fact *it completely supersedes the old Engine* . . . I believe any practical makers of machinery who would bestow sufficient time on the enquiry would arrive at the conclusion that it would be more economical to construct an engine on the new principles than to finish the one already partly

> executed and I am quite sure that one so constructed would be a much better instrument.

Having finally admitted that the first Difference Engine was obsolete, Babbage still made no specific recommendation for its abandonment. He dodged the issue, declared himself to be no more than the honest messenger of new intelligence, and haughtily left the choice to the Government:

> In making this report I wish distinctly to state that I do not entertain the slightest doubt of the success of the first Engine nor do I intend it as an application to finish the one or to construct the other: but I make it from a conviction that the information it contains ought to be communicated to those who must decide the question relative to the Calculating Machine.

The authorities could scarcely be blamed for not knowing quite what to make of Babbage's indignant pleas. No reply was made. Babbage let the matter rest for nearly two and a half years until July 1838, when he wrote to the Prime Minister, now Viscount Melbourne, in apparent desperation: 'I now appeal to your Lordship for the last time to ask for no favor [*sic*] but to ask for that which is an injustice to withhold from me – a decision.' The Chancellor, Thomas Spring-Rice, replied in a well-meaning way, saying that he was unclear what question Babbage wished him to decide, and courteously asking Babbage to make his position clear – whether he wished to finish the old or commence the new. Again Babbage bucked the issue, and repeated his request that the Government decide. Spring-Rice must have shrugged in exasperation. He did not reply, and the stalemate lasted for three years more.

The desultory exchanges with government were episodic irritations during the boom years of Babbage's work on the Analytical Engine. In December 1838 he decided to resign the Lucasian Chair of mathematics at Cambridge to devote himself even more fully to the Engine. Most of his other activities took a back seat, but his energetic social life continued unabated, and the publication of his *Ninth Bridgewater Treatise* in 1837 enhanced his standing among the intelligentsia. His theory that miracles could be seen as lawlike discontinuities pre-programmed by God into the grand scheme of things caused a respectfully frenzied stir.

There were other distractions – the law courts and the railways. Along with many of the leading astronomers of the day, he became embroiled in one of the most notorious controversies in British astronomy, acting as what we would now call an expert witness in the bitter litigation between his friend Sir James South and the celebrated instrument-makers Troughton & Simms. Sir James, who had a reputation as an unpleasant maverick, was engaged in a running feud of phenomenal viciousness over an allegedly defective telescope mounting made for him by the instrument company.

Babbage testified on his friend's behalf, and in doing so made an implacable enemy of one Reverend Richard Sheepshanks, a supporter of Edward Troughton, who resorted to bullying Babbage after the hearing. Sheepshanks was a belligerent stirrer and a close personal friend of George Biddell Airy, who later as Astronomer Royal famously denounced Babbage's Engines as useless. Babbage was convinced that it was the personal hostility of Sheepshanks that had biased Airy against his Engines. His support for the erratic Sir James laid the foundation for one of Babbage's most unworthy public denunciations of the Royal Astronomer. But this was later, when things began to fall apart.

In the late 1830s the great railway 'gauge war' was raging. One faction advocated Brunel's more widely spaced rail tracks, another favoured Stephenson's narrower gauge of 4 feet 8½ inches. The competition for the adoption of a single standard was fierce, and tribal hostilities between the West Country and the industrial interests in the north took on a religious fervour. Babbage was invited by Isambard Kingdom Brunel to offer his views as an unofficial and unpaid consultant. He rolled up his sleeves and took five months off at his own expense, using his own assistants.

A steam locomotive and a second-class carriage were placed at his disposal. He gutted the carriage and kitted it out with instruments and recording devices to log vibration, tractional force, and the trajectory of the centre of gravity of the carriage as it was pulled around curved sections of track. His graph-plotters and pen recorders used up about two miles of paper in the process, and he duly concluded that the broader gauge was safer. He put up a convincing show demonstrating his results at a decisive meeting of the directors of the Great Western Railway in January 1839, and remained convinced that he had influenced the outcome in favour of the wider gauge. Politics and economics overrode engineering, and the broad gauge ultimately yielded to the narrow. But he was gratified to be in the thick of things and had a wonderful time careering around in his own train.

One Sunday, when the tracks should have been clear, he weighted down three carriages with thirty tons of iron ballast to investigate the effect of extreme loading. The start was delayed, and in the enforced quiet, to his astonishment, he heard another train. It was Brunel in an unscheduled chuff-chuff of his own, belting along at fifty miles an hour. But for the accident of the delay they would have met head on and impacted at ninety miles an hour, with thirty tons of additional ballast into the bargain. Instead, Babbage's train was still safely in the terminus. A shaken Babbage asked Brunel what he would have done had he seen Babbage's locomotive bearing down on him. Brunel replied that he would have put on all the steam he could command to drive the opposition off the track. Babbage and Brunel, playing 'chicken' in their trains on a Sunday.

Babbage's contributions to railway technology were characteristically inventive. He recommended the use of what we now call 'black box' recorders, 'because they would become the unerring record of facts, the incorruptible witnesses of the immediate antecedents of any catastrophe'. He devised a quick-release coupling that would prevent a derailed carriage from dragging others off the rails, and at dinner one night, in the company of officers of the new railway companies, he sketched on the back of a napkin two forms of cow-catcher – contraptions fitted to the front of the locomotive to clear the line of obstacles, especially cows. After months of serious fun, he went back to the Analytical Engine.

Although he became involved in the debates of the day, his work on the Analytical Engine was still conducted in comparative solitude. His isolation was brightened in 1840 by an invitation from Giovanni Plana, an Italian mathematician, to attend a convention of Italian scientists in Turin. Babbage was highly regarded on the Continent. He was fêted during his visits, and received over a dozen memberships and honorary titles from scientific academies abroad. He paraded his titles after his name in a showy display, perhaps as a rebuke to his own country for failing to reward him in like fashion. The mathematician Augustus De Morgan admonished him for the unseemly exhibition: 'Do set [the] confounded long tail of yours to rights. I never saw such a conger of courtesy titles.' But the attention and distinctions were a solace to Babbage, who regarded himself as hard done by at home.

In Italy, Plana was eager for more details of the new Engine. Based on what he could glean from others, he observed to Babbage:

> Hitherto the *legislative* department of our analysis has been all powerful – the *executive* all feeble. Your Engine seems to give us the same control over the executive which we have hitherto only possessed over the legislative department.

Plana had captured something quite deep in his distinction between 'legislative' and 'executive' branches of analysis. The distinction is one we would now make between 'mathematics' and 'computation' – between abstract generalised laws represented by formulae on the one hand, and the stepwise rules and techniques by which specific numerical values can be found, on the other. An example of the 'legislative' part would be the theory of algebra. The 'executive' part would be the process of carrying out the arithmetical rules using actual numbers. Over two decades earlier, Babbage had stressed the important role that computation and numerical analysis would come to play in the advancement of science. In 1822 he wrote:

> I will yet venture to predict, that a time will arrive, when the accumulating labour which arises from the arithmetical application of mathematical formulae, acting as a constantly retarding force, shall ultimately impede the useful progress of the science, unless [the calculating Engine] or some equivalent method is devised for relieving it from the overwhelming incumbrance [*sic*] of numerical detail.

Babbage was delighted with Plana's distinction, and rarely missed an opportunity to repeat it.

After six years of relative isolation, Babbage chose the Turin convention to go public with the new Engine. He left for Italy in August 1840 taking with him models, drawings and notations as visual aids as well as some of Fox Talbot's 'calotypes' – a form of photographic print – for presentation to the Grand Duke of Tuscany. Professor James MacCullagh of Dublin joined Babbage on the trip and was privy to the only known seminar on the Analytical Engine. Despite the invitation, Babbage does not appear to have been part of any official programme, and conducted his sessions in his own lodgings attended by the academic elite of Italy.

On the first day, with his formulae, drawings, notations and illustrations festooned around the room, he gave an introductory outline of the Engine. He was asked to elaborate, and his colleagues pressed for deeper explanations. He basked and bloomed in the interest and attention of geometers, engineers, scientists and mathematicians. For the first time he had the opportunity to explain his private world of computing to his peers, and to hear their response. As he later recalled:

> These discussions were of great value to me in several ways. I was thus obliged to put into language the various views I had taken, and I observed the effect of my explanations on different minds. My own ideas became clearer, and I profited by many of the remarks made by my highly-gifted friends.

Babbage derived huge satisfaction from the occasion. The acknowledgement and appreciation of his colleagues was a ringing endorsement as well as a vindication of his years of solitary work. A quarter of a century later he dedicated his autobiographical work, *Passages from the Life of a Philosopher*, to the King of Italy, Victor Immanuel II, in tribute to Victor's father, Charles Albert, who had suggested the gathering in Turin of 'Italy's choicest sons'. He expressed his debt 'for the first public acknowledgement of this invention'.

Back in England, invigorated by his visit to Turin, he started work on the so-called Small Analytical Engine – an 'entry level' machine that could be constructed at reduced cost. He also began to consider how to resolve the fate of his Engines by resuming the murky negotiations with the Government that had long since lapsed. Apart from the warm glow from the intellectual hospitality of his friends, Babbage hoped that Plana would publish a report on the Analytical Engine which would serve as an independent ratification of its merits and help his case in the new negotiations.

At the same time that Babbage was preparing himself for a new approach to the Government, he was also tentatively fishing for support on the Continent. In July 1841 he wrote to Alexander von Humboldt, a renowned geologist and long-standing associate:

> This Engine is unfortunately far too much in advance of my own country to meet with the least support. I have at an expense of many thousands of pounds caused the drawings to be executed, and I have carried on experiments for its perfection. Unless however some country more enlightened than my own should take up the subject, there is no chance of that machine ever being executed during my own life, and I am even doubtful how to dispose of those drawings after its termination.

At the start of the letter Babbage referred to the new machine for the first time as the 'Analytical Engine'. It is unclear whether he was seriously inviting a foreign country to build his Engine, or whether he hoped to use the threat of such interest as leverage in the forthcoming exchanges with the Government. In any event, nothing came of these overtures.

The bleak view of his prospects for support in England confided to von Humboldt are captured in a jaundiced piece buried in an obscure paper on taxation published in 1852. In it we find an early expression of the now clichéd refrain that British exploitation of innovation compares poorly with American entrepreneurial vigour:

> Propose to an Englishman any principle, or any instrument, however admirable, and you will observe that the whole effort of the English mind is directed to find a difficulty, defect, or an impossibility in it. If you speak to him of a machine for peeling a potato, he will pronounce it impossible: if you peel a potato with it before his eyes, he will declare it useless, because it will not slice a pineapple. Impart the same principle or show the same machine to an American, or to one of our colonists, and you will observe that the whole effort of his mind is to find some new application of the principle, some new use for the instrument.

Babbage still hoped for Plana's report from the Turin meeting. But Plana was frail and burdened by his daughter's unhappy marriage, and the task fell to a young engineer, Luigi Menabrea, who had attended the Turin seminars. Menabrea, French-born, had studied mathematics and engineering at the University of Turin, and was a rising star. He achieved the rank of general in the Italian army, went into politics and became Prime Minister of the newly united Italy in 1867. During Babbage's visit to Turin, Menabrea was thirty-one.

Babbage was annoyed that Plana had defaulted on the report and caused delay. He was relying on endorsement from distinguished and independent Continental experts to support his representations to government. Menabrea's report was a long time coming, and it was not until a year later that Babbage finally saw a pre-publication draft. His confidence in Plana's support seems to have been misplaced. Although Plana's letters to Babbage were genuinely warm and full of personal goodwill, his private views on the Engine appeared damning, and it could be that opting out of the report was not a result of personal distraction. An old friend of Babbage's, one Fortunato Prandi, warned him:

> Plana will not write anything about the engine. He seems to think that you delude yourself, that the engine, if ever executed, will be a great curiosity, but perfectly useless . . . He expressed great friendship and regard for you, but this is in substance what he told me concerning the engine. I suspect that the report made of it by Menabrea has disgusted him. He says he will write to you if you write to him. Pray say nothing of this. I tell you all I hear without restraint, but you must take care that you do not compromise me.

It is possible that Plana's dim view of the Engine, which he concealed from Babbage, was responsible for the delayed publication of Menabrea's paper. The first published description of the Analytical Engine finally appeared in French in a Swiss journal in October 1842.

Chapter 7

THE ASTRONOMER ROYAL OBJECTS

What shall we do to get rid of Mr. Babbage and his calculating machine?

Sir Robert Peel, Prime Minister, 1842

If those are your views, I wish you good morning.

Charles Babbage to Robert Peel, 1842

Robert Peel, recently appointed Prime Minister, was exhausted. It was 1842, a year marred by civil insurrection, economic crises, industrial depression, starvation and violence. He wrote to his wife, Julia, of 'great rioting and confusion' and that he was 'fagged to death' with the burdens of office.

On 22 January of that year Babbage wrote to Peel. No progress had been made on the Difference Engine since 1834, and Babbage was seeking to end the uncertainty over funding and to establish whether the Government thought that he had any outstanding obligations to the project. Peel was too preoccupied with the opening of Parliament to give the matter any attention.

During the next seven months Babbage wrote three more times but had no reply. The last of these letters is dated 12

August and addressed to Sir George Clerk at the Treasury. Peel was in the thick of the August riots, and Babbage could not have chosen a worse time to press him for an answer. Babbage wrote for the fourth time in October, again with no result.

He turned for help to an old Cambridge friend, Sir William Follett, Solicitor General under Peel. Follett is described by a contemporary as 'one of the ablest advocates and acutest lawyers of the nineteenth century' and tipped to be the next Chancellor, a distinction denied him by untimely death. Babbage's unanswered letters and Follett's nudging were not without effect. Peel, backed into a corner, began to seek advice. In August 1842 he wrote to William Buckland, a geologist to whom he turned from time to time for advice on scientific matters. Peel's scepticism for Babbage's schemes is undisguised:

> What shall we do to get rid of Mr. Babbage and his calculating machine? I am perfectly convinced that every thousand pounds we should spend upon it hereafter would be throwing good money after bad. It has cost £17,000 I believe and I am told that it would cost £14, or 15,000 more to complete it. Surely if completed it would be worthless so far as science is concerned . . . It will be in my opinion a very costly toy to complete and keep in repair. If it would now calculate the quantum of benefit to be derived to science it would tender the only service I ever expect to derive from it.

Peel goes on to compare the Government's entanglement with the Engine project with his current financial nightmare – the construction of the Caledonian Canal. The canal was haemorrhaging money, and Peel was looking for ways of extricating the Government from escalating costs and interminably deferred completion. He confides to Buckland:

> I fear a reference to the Royal Society, and yet I would like to have some authority for treating this calculating machine as I should like to treat the Caledonian Canal and would have treated it but that I was told it would cost £40,000 to unravel the web that we have spent so many hundred thousand pounds in weaving.

For Peel the engine represented the further risk of open-ended expense, and he appears less concerned with the Engine's public utility, or with the grand enterprise of science, than with finding a politically defensible excuse to cut his losses.

Peel's aversion to yet another referral to the Royal Society was well founded. Each of three earlier referrals had found in Babbage's favour and had cost the Treasury dear. It was on the strength of the first Royal Society referral in 1823, when Peel was First Lord of the Treasury, that the Government had bank-rolled the Engine construction in the first place. The second referral in 1828 had decided in favour of the Treasury settling an outstanding liability of some £4,500. The third referral in 1830 for funds to construct fireproof workshops had cost the Chancellor a goodly £2,190. The total cost to government was close on £17,500. We have seen that the Royal Society committees convened to consider the Engine question were well packed with Babbage's cronies, and Peel's reluctance to submit to the inevitable outcome of a fourth referral is understandable.

His question 'What shall we do to get rid of Mr. Babbage and his calculating machine?' was perhaps only partly rhetorical – a mock lament advertising the burdens of office. But his plea to Buckland for advice was genuine enough. He forwarded the papers to Buckland, and concluded the letter with assurances that any opinion Buckland may venture would be regarded in the strictest confidence. Buckland's response is not recorded, and a fortnight later Peel was still in search of an axe. He needed an unofficial opinion, and wrote to Buckland again: 'What do men really competent to judge say of it *in private?*'

Peel despatched his Chancellor, Henry Goulburn, to find out. Goulburn was a close friend and loyal colleague of Peel. He had been Home Secretary in Peel's first administration and Peel's Chancellor of the Exchequer throughout his second. By the mid-1840s, Goulburn was Peel's longest and closest associate in public life. Goulburn was not much of a political firecracker. His speeches were said to be dull, and his personality limited, and he routinely deferred to the dashing Peel. Even as Chancellor, Goulburn was permanently overshadowed by Peel, who was the acknowledged fiscal expert. Diligent rather than brilliant and without a strong agenda of his own to skew his task, he was probably a reliable conduit of Peel's intentions in seeking what was being whispered about Babbage behind closed doors.

Goulburn sought the advice of none other than John Herschel, who had served on all three Royal Society Engine committees and had chaired the second. He was mistakenly associated in Goulburn's mind as the instigator of the venture to build the first Engine. Herschel was by now a renowned astronomer and had been rewarded four years earlier with a baronetcy during Melbourne's Whig administration on his triumphant return from his years at the Cape of Good Hope, where he had been mapping the southern-hemisphere skies. If Goulburn had been seeking someone hostile to Babbage, Herschel was a poor choice. He was, as we have seen, an intimate friend of Babbage's, and had supported the Engine project through personal encouragement, managing affairs during Babbage's travels, service on the Royal Society committees and lobbying Wellington in private interview.

Instead of writing directly to Herschel, Goulburn used George Biddell Airy, the Astronomer Royal, as an intermediary. This was the first time that Airy was officially brought into play, and his role was crucial to the fate of Babbage's Engines. The post of Astronomer Royal was the highest office in civil science, and Airy occupied the position for no less than forty-six years from 1835. He was responsible for the Royal Observatory at Greenwich, which he ran with stern efficiency, and, though not part of his official duties, he became through diligence, excellence and career acuity the *de facto* chief scientific adviser to the Government.

Airy and Babbage were as alike as chalk and cheese. Babbage went to Cambridge as the son of a wealthy banker. Airy, ten years his junior, went as a lowly 'sizar' – a kind of student servant who received tuition in exchange for college duties. Where Babbage blew his chances to win distinction in the Senate House mathematics examinations, Airy was a brilliant star. He was top of his mathematics year in all three of his undergraduate years, and graduated as senior wrangler and first Smith's Prizeman in 1823. Where Babbage failed to secure paid professional appointment, Airy was the most successful career scientist of his generation, rising from a self-motivated student to the pre-eminent consultant engineer of his age. He advanced in steady steps from a Fellow at Trinity College, Cambridge, to the Plumian Professorship in 1828 with responsibility for the Cambridge Observatory, and from there to the coveted top post of Astronomer Royal. Whereas Babbage ached for recognition, titles and civil honours and growled at their lack, Airy refused a knighthood three times, accepting only on the fourth offer. Babbage was a bon vivant with a love of dining out and socialising, and he sparkled as a host and raconteur. With his brightly coloured waistcoats he even acquired a reputation as a bit of a dandy. Airy, in the testimony of his son, Wilfrid, 'avoided dinner-parties as much as possible – they interfered too much with his work – and with the exception of scientific and official dinners he seldom dined away from home'.

The list of Airy's accomplishments is seemingly endless. He revolutionised the organisation of the Royal Observatory and was largely responsible for Greenwich being adopted as the international datum for navigation. He received several honorary doctorates, society medals and foreign honours, civil and academic. As one of the most eminent consultant engineers in the country it was rare for a major engineering project not to be referred to him in some capacity. He served on countless government commissions, including the Railway Gauge Commission. In contrast, Babbage's involvement in the gauge war was as an outsider with no official brief. Airy's published output is prolific, with no fewer than 518 published works to his name, many on astronomy and optics, others on the Roman invasion of Britain and the influence of lunar motion on the weather.

For all his distinction, the Astronomer Royal was a pragmatic and direct man of rigid personal and professional discipline, obsessed with efficiency. In a 'Personal Sketch' of his father, Wilfrid observes that 'in everything he was methodical and orderly, and he had the greatest dread of disorder'. For the Astronomer Royal it was grit rather than grace. His language does not have the flourish of Babbage's. This is true of his scientific writing, his business letters and his personal correspondence.

Again in the words of his son, 'he kept his object clearly in view, and made straight for it, aiming far more at clearness and directness than at elegance', writing with 'great ease and rapidity'. Of the countless letters from Airy to his wife, Wilfrid comments that 'they are not brilliantly written, for it was not in his nature to write for effect . . . but they are straightforward, clear and concise'.

On questions of scientific controversy Airy was forthright and even combative: 'he never hesitated to attack theories and methods he thought scientifically wrong', and in 'debate and controversy he had great self-reliance, and was absolutely fearless'. Airy was clearly a formidable opponent. He was exceptionally well placed to influence the Government. With the Royal Observatory involved daily in astronomical observation and the production of lunar tables, he was manifestly qualified to pronounce on the utility of calculating machines.

Goulburn wrote to Airy on 15 September 1842, anxious for Herschel's opinion 'as to the probable utility of continuing to expend upon [the calculating machine] the sums necessary for its perfection'. Goulburn was candid about his own position:

> My own opinion is I confess adverse to any further public expenditure on this object because I cannot anticipate from its completion any public benefit adequate to an expenditure of from thirty to forty thousand pounds and I am therefore rather disposed to give up what the Government has already expended to Mr. Babbage and to leave it to him to deal with it as may be much in accordance with his own views and means.

But he was even-handed about the possible outcome, and sought to relieve Herschel of any responsibility for a negative result:

> If [Herschel] should feel as I do that the Machine is rather to be considered as illustrating the inventive and mechanical powers of Mr. Babbage than as conducive to any great public advantage I should feel no hesitation in acting upon my own judgement when so fortified though without committing him to the decision to which I might come. If on the other hand he should be able to state that he apprehended that the completion of the Machine to its full extent was likely to be a public benefit I would then proceed to examine more minutely the expense necessary to be incurred and to reconsider my present opinion.

As well as enlisting Airy's help in securing Herschel's opinion, Goulburn concluded his letter by casually inviting Airy's views: 'My object therefore is to ascertain whether you could obtain for me Sir J. Herschel's opinion of the matter. If you could add your own also it would be conferring on me an additional favor [*sic*].'

Airy did not forward Goulburn's letter to Herschel, but transmitted the Chancellor's request in a letter of his own. But the throwaway invitation to offer his own views was irresistible. Airy wrote immediately, offering his views on the 'unfortunate business', and his views are damning. He discredits the favourable findings of the Royal Society by claiming that its specially convened committees were populated with Babbage's acolytes 'blinded by the ingenuity' of the invention. He relates that the council of the Royal Society was not in full agreement about the benefits of the machine, and that the dissenter, Dr Young, who objected, 'was regarded by Mr. Babbage with the most intense hatred'. So sensitive was Babbage that Airy and others apparently refrained for years from mentioning the Engine in his presence. Airy then moves from Babbage's irrational defence of the Engines to scotch the notion that the Engine has any general applicability:

> An absurd notion has been spread abroad, that the machine was intended for *all* calculations of every kind. This is quite wrong. The machine is intended *solely* for calculations which can be made by addition and subtraction in a particular way. This excludes all ordinary calculation.

Airy then savages the notion that the machine, capable only of specialised calculations, would be of any practical use at all:

> Scarcely a figure of the *Nautical Almanac* could be computed by it. Not a single figure of the Greenwich Observations or the great human Computations now going on could be computed by it. Indeed it was proposed only for the computation of new Tables (as Tables of Logarithms and the like), and even for these, the difficult part must be done by human computers. The necessity for such new tables does not occur, as I really believe, once in fifty years. I can therefore state without the least hesitation that I believe the machine to be useless, and that the sooner it is abandoned, the better it will be for all parties.

Airy's attack on the prospective utility of the machine is fourfold. He discredits the favourable findings of the committee on the grounds that its views were less influenced by the merits of the machine than by adulation of its inventor; he scotches the notion that the Engine has general applicability, and then denounces its prospective benefits to current computational tasks in astronomical navigation and observational astronomy. Finally, he demolishes as ill-conceived the notion that there was a significant demand for new tables at all. His rejection of the utility of the machine is robust and complete. His account makes no concessions to the finer feelings of those who supported the machine, and the brusque and even bruising tone of the letter suggests strong feeling – resentment, indignation and even anger.

The bluntness of Airy's letter is not uncharacteristic of the man's style, but even so the views he expressed are unmistakably strong. Goulburn's letter is marked 'Private and Confidential' and all subsequent exchanges are headed 'Private' by both parties. With this protection it may seem that Airy gave voice to views that might otherwise have been moderated by propriety – even by the possibility of disclosure to Babbage himself. Quite the reverse. He freed Goulburn from any constraints, and gave the Chancellor licence to use the report as freely as he wished. He wrote to Goulburn that he was 'fully entitled, in courtesy as well as in right, to use my expressed opinion in any way that you shall think fit'.

Goulburn was evidently gratified by Airy's response: Airy's condemnation confirmed Peel's view that the Engine was a 'very costly toy' and fortified Goulburn's own reservations. He wrote to Airy by return, tendering his 'best thanks for your satisfactory letter'. The most eminent judge in the land had pronounced on the benefits of Babbage's calculating machine, and the verdict was damning.

Airy was drawn into the Engine affair late in the day – twenty years after the genesis episode when Babbage and Herschel were checking tables, and about ten years after the construction project was abandoned. But his opposition to the Engines was well known much earlier, and he had not been coy in expressing his views in private. An Irish astronomer, Thomas Romney Robinson, writing to Babbage from Armagh Observatory in 1835, gave some idea of the intrigues surrounding Babbage and Airy's opposition:

> The opinion which I (and Beaufort simultaneously) formed respecting Airy's being privy to some plan of attacking you, arose from manner and look [rather] than anything which he actually said. Indeed of the conversation I distinctly remember but two points, one his saying 'that the persons who had recommended the construction of the machine would shortly find themselves in a very unpleasant predicament' the other 'that in his opinion the machine was useless, for that if the money spent on it had been applied to pay computers, we could have had all that is wanting in the way of tables'. Airy I do not think is likely to have been the mover in this, but wherever it comes from, let me entreat you not to despise the attack as unimportant because it is contemptible.

Airy's scepticism appears to have been known very early. Later in the same letter Robinson, perhaps as a form of moral support, portrays Airy as isolated in his opposition to the machine: 'If [Members of Parliament] go into the merits or use of the invention itself I cannot suppose they will find a man in Europe but Airy to gainsay it.'

Airy was not alone in his less than absolute commitment to the utility of the Engines. Others raised questions about whether the high precision offered by the machines in working to twenty, forty, fifty and even one hundred figures of accuracy could be justified when practical measurements could be made to only three or four figures. There were also those who asserted that conventional methods of producing tables were sufficiently reliable, though this was difficult to verify. Experts disagreed, and among the coterie of doubters Airy's views were evidently widely known. In a piece of contemporary social reportage the famous actor William Macready, whose diaries provide an inexhaustible source of society gossip, made an entry in September 1837 on Babbage's Engine: 'Professor Airy says the thing is a humbug. Other scientific men say directly the contrary.'

Babbage was convinced that Airy was jealous of him. The question of whether any personal antagonism towards Babbage prejudiced Airy's professional judgement against the Engines remains open. He was certainly a diligent and conscientious civil servant, and it would be uncharacteristic of him to have allowed his personal feelings to compromise the soundness of findings made in his official capacity. But there is an early episode that may just contain the seeds of later rivalry. As an undergraduate at Cambridge in 1822, Airy had read Babbage's short notice, just published, which announced the invention, and he was stimulated to sketch a machine of his own for solving equations. In later life Airy was offhand about his interest in the machine, but William Whewell, a Cambridge don writing at the time, indicates otherwise. In October 1822 he wrote to Herschel:

> You have of course heard from [George] Peacock about Airy, a pupil of his, and certainly a man of very extraordinary talents. The reports about Babbage's machine have, it seems, excited him to attempt something of the same kind. He and another man have made a machine to solve cubic equations, but besides this he has, so far as I can make out, invented a good deal in the way of Babbage's contrivance. He is not here at present, but his friend tells me that it has got

> toothed wheels, working in one way for the differences and in another for the digits. If it be a similar invention to that, it is probably an independent one; for I do not know any way by which he has got any lights about Babbage's affair.

It is tempting to cast Airy as the Government's hit-man – someone whose known opposition to the case would produce the desired official condemnation. But Goulburn was writing to Airy to solicit Herschel's views, and his invitation to Airy to express his opinion, at least at face value, was a casual afterthought.

Herschel in the meantime was squirming over his brief. It was a sensitive business, and he sent his report to Airy sealed to ensure 'perfect insulation of minions', for onward transmission to Goulburn. The report ended ten days of flurried exchanges. Herschel's report is long and elaborate. In it he steers an uncomfortable path between loyalty to Babbage, an apparently genuine belief in the utility of the machine, and public duty. The document is clearly well meant, but the embroidery and circumspection weaken his stated conviction in the benefits of the machine. If Goulburn's subsequent action was influenced by either of the two reports, then Airy's short clear damnation carried more weight than Herschel's carefully considered support. Goulburn wrote to Babbage on 3 November 1842, axing the project on the grounds of cost.

In his posthumous 'Autobiography' compiled by his son, Airy recounted the episode of his role in the demise of Babbage's Engine as follows:

> On Sept. 15th Mr. Goulburn, Chancellor of the Exchequer, asked my opinion on the utility of Babbage's calculating machine, and the propriety of

> expending further sums of money on it. I replied, entering fully into the matter, and giving my opinion that it was worthless. – I was elected an Honorary Member of the Institution of Civil Engineers, London.

This is the most frequently quoted of Airy's condemnations. It despatches half a lifetime's work in a brusque dismissal, with no trace of supportive argument or self-justification. The casual mention of his election to the Institution of Civil Engineers adds insult to injury, seemingly ranking his honorary membership of a professional institution with Babbage's grand venture. We should not read too much into the apparent brutality of the entry. The 'Autobiography' has the format of an annual diary, and there are many other abrupt transitions between unrelated events. The record for 1826, for example, reconstructed from Airy's personal papers, memorabilia and the scratch-pads used for daily workings, juxtaposes a record of his stipend as a tutor (£50) and the ellipticity of heterogeneous spheroids. A starker non sequitur from the same year places Airy's commencement of tuition in Italian alongside his attendance of the guillotining of a murderer in the Place Martroi, and investigations into pendulums and the calculus of variations.

Goulburn's letter to Babbage dated 3 November 1842 was the end of the line for the Difference Engine, or so it seemed.

> We both [Goulburn and the Prime Minister, Sir Robert Peel] regret the necessity of abandoning the completion of a Machine on which so much scientific ingenuity and labour have been bestowed. But on the other hand, the expense which would be necessary in order to render it either satisfactory to yourself, or

> generally useful, appears on the lowest calculation so far to exceed what we should be justified in incurring, that we consider ourselves as having no other alternative. We trust that by withdrawing all claim on the part of the Government to the machine as at present constructed and placing it at your entire disposal we may so assist afresh in your future exertions in the cause of science.

After twenty years the Treasury was axing the project, writing off the massive expense and offering Babbage the physical debris of his labours for his own use at no charge. With Babbage's touchpaper sensitivities and his history of intemperate behaviour, we might be forgiven for expecting an eruption of some kind – despair, anger or outrage. But events took an unlikely turn. Babbage wrote a calm note to Goulburn on 6 November in which he declined the gift of the Engine on the grounds that he had throughout regarded 'the drawings and the parts of the machine already executed' as the absolute property of the Government and entertained no claim to them. On the same day he wrote separately to Peel as the joint author of the decision.

The letter is considered and courteous. He repeats that he is compelled to decline the gift of the Engine, and adds:

> I infer however both from the regret with which you have arrived at the conclusion as well as from the offer itself that you would much more willingly assist at the creation of the Analytical Engine than become the official cause of its total suppression or possibly of its first appearance in a foreign land . . .

The inference of Peel's secret wishes is, on the face of it, extraordinary. Goulburn's letter contained no reference at all to the Analytical Engine. Babbage continues:

> I also perceive in the expression of feeling that although I have no other than a moral claim for twenty years of exertion accompanied by great pecuniary as well as personal sacrifices yet those exertions in the cause of Science ought not to remain utterly unrecognised and unrequited by the Government.

The only part of Goulburn's letter that might conceivably have suggested anything of the kind is where he expresses his regret at the abandonment of a machine 'on which so much scientific ingenuity and labour has been bestowed'. Goulburn's phrase was more than likely a politeness to cushion disappointment. For Babbage to construe an innocent tribute as an unexpressed recognition by the Government that he was somehow entitled to reward appears only slightly less wishful than his construal of Peel's secret wish to start funding the Analytical Engine. After the expensive embarrassment of the Difference Engine No. 1, the idea would surely have appalled them. With a courteous flourish (that it would be unjust to both of them to seek a decision without personal communication), Babbage said that he would esteem it a favour if Peel would allow him an interview.

Taken at face value, Babbage's response is puzzling. It is less strange when seen as part of a deeper strategy revealed from surviving notes he made in preparation for the meeting with Peel. It seems that Babbage's repeated insistence on a decision from the Treasury during his years of barracking was not, as has been supposed, in order to salvage renewed support for the Difference Engine, but in fact to get it scrapped. Once the

Treasury had axed the project, he would argue that the Government had an obligation to him – first for the wasted years which were deserving of reward, and secondly for reneging on the original deal to fund an engine to completion. With the Government impaled (imagined Babbage) on its own awful sins, and therefore desperate to right these grievous wrongs, he would offer them salvation – the opportunity to build a small Analytical Engine or simplified Difference Engine for less than it would cost to complete the old Difference Engine No. 1. Indeed, in April of that year he had already begun to sketch out a new and elegantly simple Difference Engine No. 2.

This interpretation of Babbage's strategy answers the puzzle of why he did nothing to argue for continuance in his repeated insistence that it was for the Government to decide whether it wished him to complete or abandon the project. Only once the first Difference Engine was scrapped could Babbage point to the Government's debt to him and implement the second phase of the strategy, a new proposal on a fresh financial basis, plucked like a rabbit out of the magician's hat.

The surviving agenda notes give clues as to how Babbage proposed to steer the meeting with Peel. The first element related to his grievance about being entitled to reward. He cites the injustice of the personal sacrifices he had made in refusing lucrative employment, including following his father into City banking, and in being constantly in debt to his engineers for long periods of time while the moneys were recouped from the Treasury. His notes go so far as to list, rather pitifully, the salaries, pensions (which were then discretionary) and benefits conferred on other scientists, Airy, Herschel and Whewell included – rewards that had been denied him. The second element of the agenda notes is a financial programme that would allow him to complete the

Top: Charles Babbage and fiancée Georgiana Whitmore, 1813. Miniatures from engagement locket.
Bottom: John Herschel (1792–1871) in 1836. Astronomer, philosopher and pioneer of photography. Babbage's lifelong friend.

4

Log.	N.	Log.	N.	Log.	N.	
. 865 6960 599	801	2. 903 6325 161	867	2. 938 0190 975	934	2. 970
. 866 2873 391	802	2. 904 1743 683	868	2. 938 5197 252	935	2. 970
. 866 8778 143	803	2. 904 7155 453	869	2. 939 0197 765	936	2. 971
. 867 4674 879	804	2. 905 2560 487	870	2. 939 5192 526	937	2. 971
. 868 0563 618	805	2. 905 7958 804	871	2. 940 0181 550	938	2. 972
. 868 6444 384	806	2. 906 3350 418	872	2. 940 5164 849	939	2. 972
. 869 2317 197	807	2. 906 8735 347	873	2. 941 0142 437	940	2. 973
. 869 8182 080	808	2. 907 4113 608	874	2. 941 5114 326	941	2. 973
. 870 4039 053	809	2. 907 9485 216	875	2. 942 0080 530	942	2. 974
. 870 9888 138	810	2. 908 4850 189	876	2. 942 5041 062	943	2. 974
. 871 5729 355	811	2. 909 0208 542	877	2. 942 9995 934	944	2. 974
. 872 1562 727	812	2. 909 5560 292	878	2. 943 4945 159	945	2. 975
. 872 7388 275	813	2. 910 0905 456	879	2. 943 9888 751	946	2. 975
. 873 3206 018	814	2. 910 6244 049	880	2. 944 4826 722	947	2. 976
. 873 9015 979	815	2. 911 1576 087	881	2. 944 9759 084	948	2. 976
[illegible]	816	[illegible]	[illegible]	[illegible]	[illegible]	[illegible]

5

Top: Excerpt from a page of logarithm tables by Georg von Vega, 1794.
Bottom: Thesaurus Logarithmorum Completus by Georg von Vega, 1794.

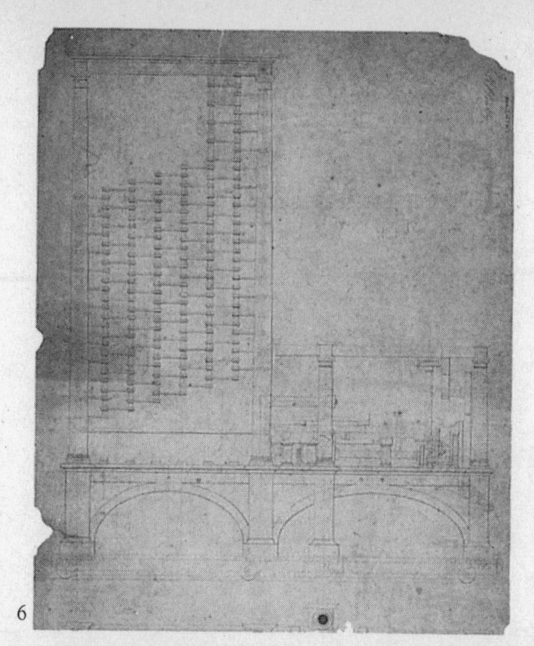
6

7

Top: Design drawing, Difference Engine No. 1, 1830.
Bottom: Difference Engine No. 1 – portion, 1832.

8

9

Top: Ada, Countess of Lovelace, Lord Byron's daughter, by A. E. Chalon, circa 1838.
Bottom: Babbage's house, 1 Dorset Street, Marylebone. Sketch by Hanslip Fletcher.

10

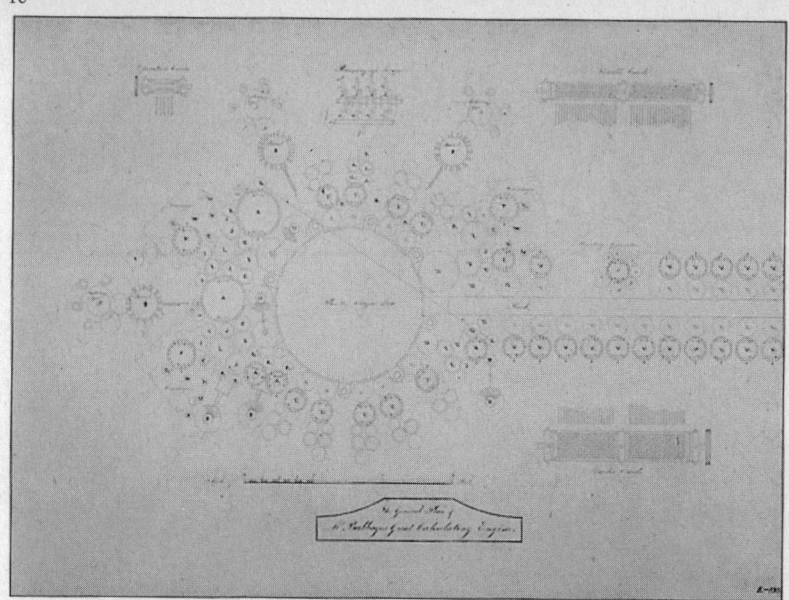

11

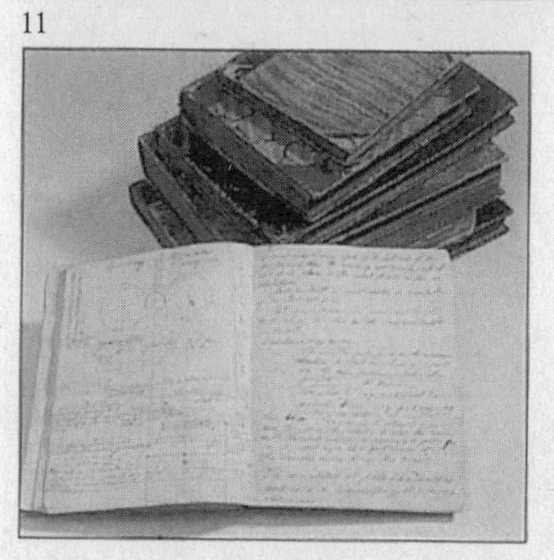

12

Top: Design drawing, Analytical Engine, 1840. The Mill ('central processor') is distributed around the large central circle. The Store ('memory') is to the right and shows nineteen registers.

Bottom left: Scribbling Book, 1836.

Bottom right: Punched cards for the Analytical Engine. The smaller Operation cards specify the arithmetic operation. The larger Variable Cards indicate the location of data in the Store.

13

14

15

Top left: The first Scheutz difference engine, Sweden, 1843.
Top right: Georg Scheutz (1785–1873).
Bottom: The third Scheutz difference engine, 1859.

16

17

Top: George Biddell Airy (1801–1892) in 1833–4.
Bottom: Babbage (1791–1871), circa 1849. Daguerreotype by Antoine Claudet.

18

19

Top: Lady King (*née* Augusta Ada Byron), by Margaret Carpenter, 1835.
Bottom: The Great Exhibition, Crystal Palace, 1851. The Transept looking north.

20

21

22

Top: Babbage's shoes for walking on water.
Bottom left: A Jacquard Loom.
Bottom right: Portrait of J. M. Jacquard woven in silk, circa 1839.

23

24

25

Top left: Mill of the Analytical Engine – detail, 1871.
Top right: Doron Swade, 1995.
Bottom: Mill of the Analytical Engine – portion, 1871.

26 27 28 29

Top left: Allan Bromley, London, January 1990.
Top right: Reg Crick, London, January 1990.
Bottom left: Michael Wright, London, January 1990.
Bottom right: Barrie Holloway, London, Spring 1991.

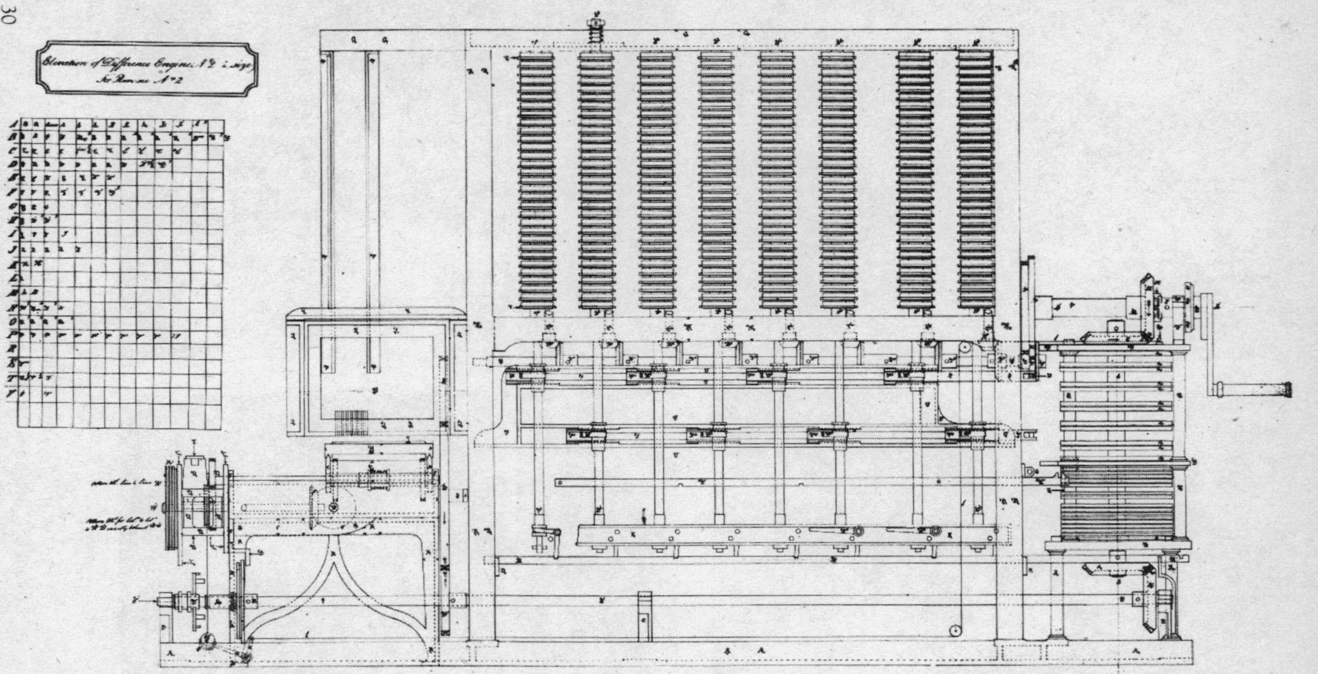

Design drawing, Difference Engine No. 2, 1847. The cam stack and crank-handle are on the right, eight columns of figure wheels in the centre, and the printing and stereotyping apparatus are shown on the left. The matrix (far left) is an example of Babbage's Mechanical Notation.

31

32

33

Top left: Trial piece for Difference Engine No. 2, 1989.
Top right: Difference Engine No. 2, calculating section, May 1991.
Bottom: Engineers in the 'cattle pen'. Barrie Holloway (left), Reg Crick (right).

34

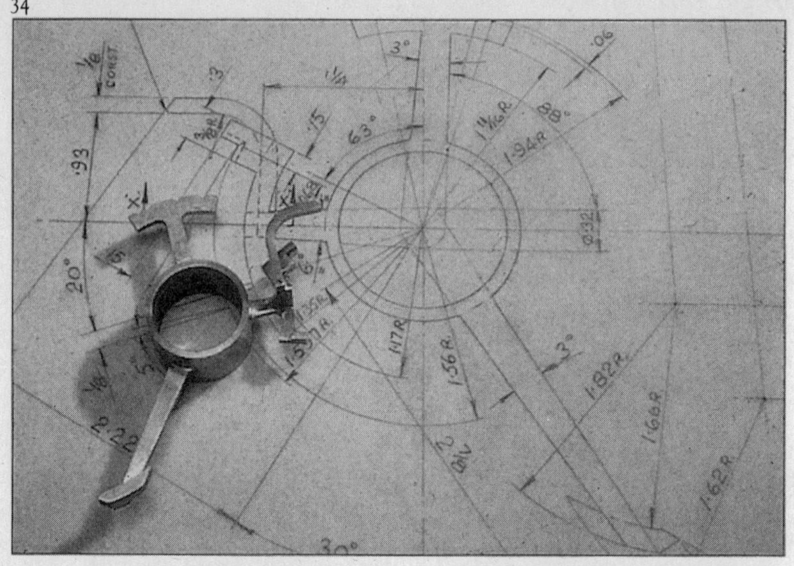

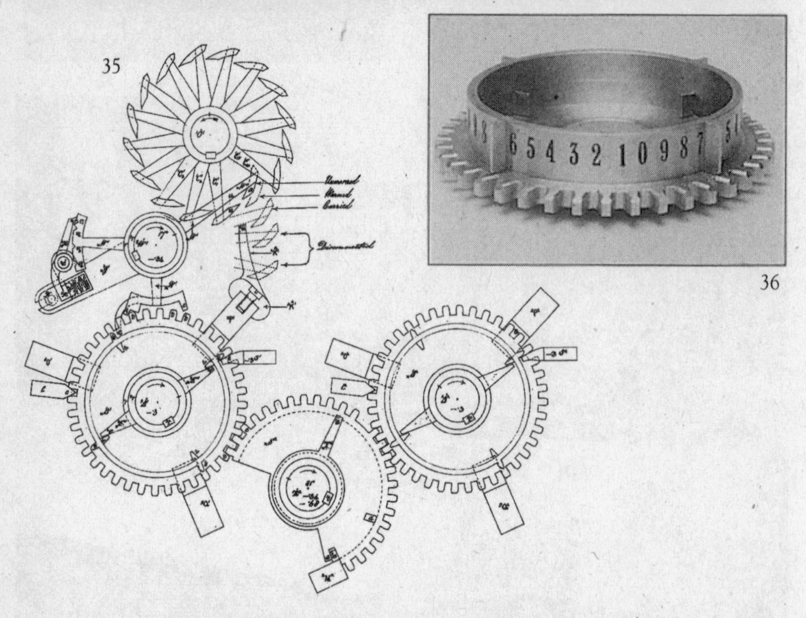

Top: Levers for the carriage mechanism shown against a modern fully-dimensioned drawing.
Bottom left: Design drawing, Difference Engine No. 2 – detail, 1847. Section through mechanism for addition and for carrying tens.
Bottom right: A figure wheel – one of 248

37

38

39

Top: Difference Engine No.2. Designed 1847–9. Calculating section (rear) completed 1991; printing and stereotyping apparatus (front) completed 2001.
Bottom left: Mill of the Analytical Engine by Henry Prevost Babbage, 1910.
Bottom right: Major-General Henry Prevost Babbage (1824–1918).

40

41

42

Three of the main players in old age.
Top: John Herschel (1792–1871) in 1867.
Middle: George Biddell Airy (1801–1892).
Bottom: Charles Babbage (1791–1871) in 1860.

Difference Engine. This involved three equal stages of payment, one as an advance, the second when the Engine was half-finished, and the third on the successful printing of sample logarithm tables. He also proposed that the copyright of all mathematical tables, tables of interest and tables used in all ready-reckoners reside with him as a source of income. As an alternative he proposed that he should be appointed to a paid post that would allow him to finance the machine from his own pocket.

Such is the reconstruction of Babbage's strategy. In anticipation of a meeting with Peel, Sir James South, Babbage's maverick friend, tried to warn Babbage that Peel was beleaguered and that Babbage should tread warily:

> I write this to *entreat* you in your interview with Peel which we think he will not deny you, not to say anything which *can possibly* irritate him. He is at present in a false or bad position, pray mind *you* do not change places with him.

Peel gave Babbage one day's notice of the appointment: 'Sir Robert Peel presents his compliments to Mr. Babbage and will have the pleasure of seeing him at Eleven o'clock tomorrow (Friday) morning.' The meeting was set for 11 November 1842 at Whitehall. Sir James's warning was written on the day of the meeting. If Babbage received it in time, he paid it no heed.

The meeting misfired catastrophically. Babbage profoundly misjudged Peel's mood. Peel, in office just over a year, was exhausted by one of the most politically taxing years of the century. There was civil unrest and he was having to deal with constant financial crises. He was also thoroughly fed up with the endless submissions from those hopeful of patronage, honours and financial rewards from the new administration. He had sarcastically complained that 'he had not had a single application for office from anyone fit for it', and not having a knighthood or civil honour was becoming a rare distinction. Someone less proud, principled and aggrieved than Babbage might have read the signs and succeeded in prevailing upon Peel who was, publicly at least, well disposed towards science. Not Babbage.

He bounded in and launched into phase one of the prepared strategy – to establish the Government's debt to him. He regaled Peel with his list of injustices. He insinuated that Peel's scientific advisers who had counselled against the Engines were motivated by professional jealousy. He laboured his personal and financial sacrifices, and laid moral claim to some compensatory reward since it was the Government's decision, not his, to abandon the project. He also vented his sense of injury at public allegations that he had profited personally from government funds, and cited instances in which other distinguished scientists had been favoured by pensions, emoluments and other benefits. He implied that a civil honour would rectify any public misconception of wrongdoing.

Peel was having none of it. He countered by telling Babbage that he had rendered the Difference Engine useless by inventing a better one, and denied that Babbage was owed anything. Peel argued that the awards to other scientists were in the way of rewards for professional services, not sinecure payments. Babbage objected. His long account of the meeting, written on the same day, bristles with the detail of immediate recall:

> Sir Robert Peel seemed excessively angry and annoyed during the whole interview but more particularly when I knocked over with some vivacity his argument about

> *professional service* . . . I listened to all his statements looking him steadfastly in the face. When he got aground, I still retained my view upon him as if expecting at least some argument would be produced. This position of course was not very agreeable and certainly not very dignified for a prime minister.

Babbage was furious. 'If those are your views', he said, addressing Peel, 'I wish you good morning.' He turned on his heel and stormed out. Staring down an already ratty PM was not the best tactic to secure anything. His feelings had again got the better of him; protest had triumphed over persuasion. He never did get to put forward his new proposal. With his dramatic exit went the last serious prospect of completing any of his machines.

In the story of Babbage's Engines, history portrays Airy as an unimaginative bureaucrat – a mediocre but influential insider playing Salieri to Babbage's Mozart. When we trace the source of these perceptions it emerges that they all derive from none other than Babbage. Without exception the contemporary accounts of the doomed fate of Babbage's Engines are based on sources provided personally by Babbage. He is a vigorous and articulate writer, and his historical voice, loud and strong to begin with, is amplified by the particular soft spot history seems to have for thwarted geniuses, for visionaries and for men 'ahead of their time'. Babbage was a prominent gentleman, Airy a civil servant, and Airy's voice has been largely ignored except to reinforce stereotypes of bureaucratic dullness.

The dim view of Airy taken by history is not helped by the relentless ignominy he suffered by not discovering the planet Neptune. The discovery of a new planet was of huge scientific and public interest, especially to astrology which was even more popular then than it is now. A young Cambridge astronomer, John Couch Adams, concluded after laborious calculation that the irregularity of the orbit of Uranus might indicate the existence of a new planet, but Airy chose not to conduct a telescopic search. The planet was found by Johann Galle observing from Berlin in 1846, after a search prompted by a memoir from the French astronomer Urbain Le Verrier. The scientific world and the British public were unforgiving, and Airy's lapse, though professionally defensible, was seen as a serious failure of imagination. He was vilified for years after. Airy's blunder over Neptune has helped Babbage historians to promote the Astronomer Royal's role as that of an uninspired bureaucrat and allows them to collude in the portrayal of Babbage as a genius surrounded by fools.

The demise of the Engine project was the central trauma in Babbage's professional life. He lived and relived the sorry circumstances until his death. Time and again in his writing he replays these events, sometimes in outrage, sometimes in protest, disdain, incomprehension, self-justification or despair, as though unable to reconcile himself to the dismal outcome.

Chapter 8

THE ENCHANTRESS OF NUMBER

The more I study, the more insatiable do I feel my genius for it to be.

Ada Lovelace, 1843

Augusta Ada, Countess of Lovelace, now began to play a role in Babbage's battered aspirations for his Engines. Ada, only legitimate daughter of the poet Lord Byron, first met Babbage at a party in June 1833 when she was seventeen. Lady Byron, Ada's mother, describes the occasion in a letter:

> Ada was more pleased with a party she was at on Wednesday . . . she met there a few scientific people – amongst them Babbage, with whom she was delighted . . . Babbage was full of animation – and talked of his wonderful machine (which he is to shew us) as a child does of its plaything . . .

Twelve days later Babbage demonstrated the machine to mother and daughter, and Ada was entranced by its workings and its enthusiastic demonstrator. 'We both went to see the *thinking* machine,' wrote Lady Byron, 'for such it seems.' Ada had rank, beauty and an interest in mathematics. As Byron's daughter, she was also an object of public fascination and became an attractive young addition to Babbage's soirées.

Byron had rocketed to fame with the publication in 1812 of *Childe Harold's Pilgrimage*, a romantic epic based on his time in Greece. Fame, rank, a philandering lifestyle and a limp made for a cocktail that intoxicated his public. He was famously described for his drunken excesses and general debauchery as 'mad, bad, and dangerous to know'. Byron married Annabella Milbanke in January 1815. Their relationship has been described by one of Byron's biographers as 'one of the most infamously wretched marriages in history'. Ada was born in December of that year. Shortly after, Byron attempted to rape his wife on four occasions, and servants resorted to locking her door against him. Drunken abuse and an attempt by Byron to evict his wife followed, and within a few weeks of Ada's birth Annabella had fled the marital home, taking her infant daughter with her. Annabella sought a separation amidst rumours of Byron's incestuous relationship with his half-sister Augusta. The rift was a public scandal, though it was said that nothing became Annabella like her separation. A settlement was reached, and Byron left England for the Continent in April 1816, leaving Ada in the care of her mother. Ada did not see Byron again. From the start her father's fame and his profligate ways made Ada a celebrity, and public interest in her was unrelenting.

Lady Byron had mathematical training (Byron had called her his 'Princess of Parallelograms') and insisted that Ada, who was tutored privately, study mathematics too. Mathematics education was unusual for women, whose constitution was regarded as too frail to withstand the rigours of its intellectual demands. Women were debarred from taking university degrees and from membership of the Royal Society. They were educated into duty and domesticity, and aristocratic women in particular were groomed for social ornamentation and household management. Science and discovery were men's work. Ada was a rebellious child, and Annabella, a disciplinarian. Annabella used mathematics as an instrument of reform.

Young Ada married well. In July 1835, at nineteen, she was wed to William King, the eighth Lord Ockham. Several honours followed in brisk succession. In Queen Victoria's coronation list King was elevated to an earldom and became the first Earl of Lovelace in 1838. In 1840 he was appointed Lord Lieutenant of Surrey, a prestigious position in English county society, and a year later he became a Fellow of the Royal Society. William was devoted to Ada, whose interests and activities he supported and indulged.

Between 1836 and 1839 Ada, never physically strong, had three children. She was already the mistress of three family households – Ockham Park in Surrey, Ashley Combe in Somerset and a leased house in a fashionable part of London – and now had the additional responsibilities of motherhood. With domestic duties and erratic health she had little time for mathematics or for Babbage's Engines. But in 1839 she wrote to Babbage asking him to recommend a maths tutor. The services of Augustus De Morgan were secured, and Ada's studies resumed. De Morgan was a patient and gifted teacher, and Ada an enthusiastic and adventurous student. Together they worked their way through the rudiments of mathematics.

Ada became thoroughly excited by the potential of her own genius, and her letters display signs of youthful self-conceit. She wrote to her mother in February 1841:

And now I must tell you *what* my opinion of my own mind & powers is exactly; – the result of a most accurate study of *myself* with a view to my future plans, during many months. I believe myself to possess a most singular combination of qualities exactly fitted to make me *pre-eminently* a discoverer of the *hidden realities* of nature. *You* will not mistake this assertion either for a wild enthusiasm, or for the result of any disposition to *self-exaltation*. On the contrary, the belief has been forced upon me, & most slow have I been to admit it even. And now I will mention the three remarkable faculties in me, which united, ought (all in good time) to make me see *anything*, that a being not actually *dead*, can see & know . . .

Firstly: Owing to some peculiarity in my nervous system, I have *perceptions* of some things, which no one else has; or at least very few, if any. This faculty may be designated in me as a singular *tact*, or some might say an *intuitive* perception of hidden things; – that is of things hidden from eyes, ears & the ordinary senses . . . This *alone* would advantage me little, in the discovery line, but there is

Secondly; – my immense reasoning faculties;

Thirdly; my concentrative faculty, by which I mean the power not only of throwing my whole energy & existence into whatever I choose, but also bringing to bear on any one subject or idea, a vast apparatus from all sorts of apparently irrelevant & extraneous sources. I can throw *rays* from every quarter of the universe into *one* vast focus . . . Well, here I have written, what most people would call a remarkably *mad* letter, & yet certainly one of the most logical, sober-minded, cool, pieces of composition, (*I* believe), that I ever penned;

> the result of much accurate, matter-of-fact, reflection & study . . .

Ada was given to forceful emphasis and narcissism in equal measure. Some have seen her motivation and the intensity of her wish to apply herself to some great object as an attempt to atone for the 'misused genius' of her father.

A few weeks earlier she had written to Babbage placing herself at his service for some non-specific future collaboration:

> I am very anxious to talk to you. I will give you a hint on *what*. It strikes me that at some future time, (it might be even within 3 or 4 years, or it might be *many* years hence), *my head* may be made by you subservient to some of *your* purposes & plans. If so, *if* ever I could be worthy or capable of being *used* by you, my head will be yours. And it is on this that I wish to speak most seriously to you. You have always been a kind and real & most invaluable friend to *me*; & I would that I could in any way repay it, though I scarcely dare to exalt myself as to hope however humbly, that I can ever be intellectually worthy to attempt serving *you*.

She directed her mathematical studies to the Difference Engine: 'I am now studying attentively the *Finite Differences* . . . and in this I have more particular interest, because I know it bears directly on some of *your* business.' Babbage, not long back from his successful trip to Turin, was preparing to renew the correspondence with the Treasury in the hope of renegotiating the financial basis on which the Engines might be built. Her interest in his work must have been welcome in a world which had seemed indifferent to his efforts. Babbage became a family friend and was a frequent visitor at the Lovelaces' homes.

Babbage's hopes for his Engines, revived by his successful trip to Turin and buoyed by the novelty of Ada's interest, were dashed by the bruising encounter with Peel. There was now no prospect of further funding from the Government, and Babbage gave up any immediate hopes of constructing any of the Engines, even a simplified Difference Engine. The Duke of Somerset, Babbage's close friend and frequent dining companion, pressed Babbage to visit his family in Devon in the hope that the trip would cheer him up.

In October 1842, a few weeks before the debacle at Whitehall, the description of the Analytical Engine prepared by the Italian engineer Luigi Menabrea finally appeared in a Swiss journal. It had taken just over two years from Babbage's lectures in Turin to reach print. The article, 'Notions sur la machine analytique', was perhaps some small comfort to Babbage in the weeks following the encounter with Peel.

On the suggestion of Charles Wheatstone, a scientist and family friend, Ada translated Menabrea's paper from the original French into English for a published digest, *Scientific Memoirs*, which specialised in foreign scientific papers. Ada's French was proficient, and under Wheatstone's supervision the translation was duly done. It seems that Babbage was ill in the autumn of that year, and the translation was presented to him as a fait accompli early in 1843. He recalls the event some twenty years later in his autobiography, *Passages*:

> Some time after the appearance of [Menabrea's] memoir . . . the late Countess of Lovelace informed me that she had translated the memoir of Menabrea. I asked why she had not herself written an original paper on a subject with which she was so intimately

> acquainted? To this Lady Lovelace replied that the thought had not occurred to her. I then suggested that she should add some notes to Menabrea's memoir; an idea which was immediately adopted.

Ada, unleashed, threw herself into the expansion of Menabrea's article, and a frenetic collaboration with Babbage followed. Letters, notes and messages flew between them, and consultations in London were frequent. Ada worked with frantic energy. She became demanding, bossy, coquettish and irritable. She badgered Babbage for explanations about the operation of the Analytical Engine, reprimanded him for carelessness in mixing up her drafts, and gave him stern instructions, threatening him with her annoyance if he did not comply.

For Babbage the major work on the Analytical Engine was already done. He was now tinkering with improvements without any great sense of direction. His Engine was unbuilt, Peel had rebuffed him, and his country had largely ignored his efforts. Here was someone consumed with the great importance of his work. He was surely gratified and charmed by Ada's fervour, especially after the hurtful indifference of his contemporaries and his practical disappointments. Babbage supplied her with the examples he had used in Turin which he had drawn from his work years earlier. He was encouraging, friendly and flattering, correcting her misapprehensions, annotating the drafts and guiding the writing.

As her notes neared their final form, Ada became elated at her own prowess. She wrote to Babbage at the end of July 1843, 'the more I study, the more insatiable do I feel my genius for it to be'. She also compared her talents to those of her poet-father: 'I do *not* believe that my father was (or ever could have been) such a *Poet* as *I shall* be an *Analyst* (&

Metaphysician).' She took to signing herself as Babbage's 'Fairy'. A day later she was manic with self-regard and wrote again to Babbage, referring to the article on the Engine as her 'child':

> I cannot refrain from expressing my amazement at my own child. The *pithy & vigorous* nature of the style seem to me to be most striking; and there is at times a *half-satirical & humorous dryness*, which would I suspect make me a most formidable reviewer. I am quite thunder-struck at the *power* of the writing.

But all did not go well. In Ada's additions to Menabrea's article she attempted to distance herself from the politics of Babbage's grievances against the Government by excluding any reference to the history of the construction project and the sorry outcome. With the publisher's deadline looming, Babbage drafted a self-justifying statement recounting the circumstances of the Engine projects and his disputes with various governments. His intention was that this should accompany Ada's translation and notes in the form of an anonymous addition, so masking its provenance. Babbage was deputed by Ada to deliver her manuscript to the publishers so she was not privy to Babbage's negotiations for his addition. The publishers were reluctant to include the statement, and Babbage asked Ada to withdraw the article and publish it elsewhere.

Ada was livid. Babbage had assumed throughout that in expanding the article she was acting solely in his service, as she had pledged to do, and that she would automatically comply with his wishes. She, on the other hand, had long since taken it for granted that the article was hers and that Babbage was her assistant to be commanded by her in the realisation of a great destiny. After all her work, her conviction in her genius, here she was on the brink of greatness, only to find Babbage hijacking her masterpiece as a vehicle for his sordid squabble with the authorities. She would have none of it and told Babbage so. The same day she wrote to her mother with her account of the quarrel:

> I have been harassed and pressed in a most perplexing manner by the conduct of Mr. Babbage. We are in fact *at issue*: & I am sorry to have come to the conclusion that he is one of the most *impracticable*, *selfish*, & *intemperate* persons one can have to do with . . . I declared at once to Babbage, that no power should induce me to lend myself to any of his quarrels, or to become in any way his *organ*; & that I should myself communicate in a direct manner with the editors on the subject, as I did not choose to commit a dishonourable breach of engagement, even to promote *his* advantage . . . He was *furious*; I am imperturbable & unmoved. He will never forgive me.

Babbage retired hurt. Ada's article went off to the printers in August 1843 and appeared in Richard Taylor's *Scientific Memoirs* without the offending statement. The author of the published article was identified only by Ada's initials, 'A.A.L.', as it was not the done thing for a woman, particularly one of high rank, to put her name to a scientific paper. Babbage published his 'statement' anonymously in the *Philosophical Magazine* a few weeks later.

While relations were strained, Ada wrote Babbage a marathon twelve-page letter of self-justification. In it she stipulated a series of conditions that Babbage must comply with to secure her continued efforts on his behalf. Ada proposed to 'front' Engine construction and so leave Babbage to concentrate on technical issues. Babbage was to agree to abide wholly by her judgement, or that of named referees, in all matters of negotiation and public relations. She charged Babbage to

> undertake to give your mind *wholly* & *undividedly* . . . to the consideration of all those matters in which I shall at times require your intellectual *assistance* & *supervision* & can you promise not to *slur* & *hurry* things over; or to mislay, & allow confusion & mistakes to enter into documents etc.

Babbage was no great diplomat. In fact, when it came to his relationships with governmental and scientific organisations there are distinct signs of ineptness. Ada was evidently aware of this and was, in the glow of her publication triumph, grandiosely offering Babbage her prized services to provide a protective buffer between him and the world. The intention may have been sound, but the tone and the quasi-legal formulation of the proposal and conditions are curious. Babbage rejected the overture out of hand.

The rift was short-lived. On 22 August Ada wrote to Lady Byron, 'Babbage & I are I think more friends than ever. I have never seen him so agreeable, so reasonable, or in such good spirits!' Ada pressed Babbage to visit the Lovelaces at Ashley Combe. Babbage was evidently relieved that the breach was not final, and was carefree in his reply:

> I find it quite in vain to wait until I have leisure so I have resolved that I will leave all other things undone and set out for Ashley taking with me papers enough to enable me to forget this world and all its troubles

> and if possible its multitudinous Charlatans –
> everything in short but the Enchantress of Number.

It is not obvious who or what is referred to by the phrase 'Enchantress of Number'. In the interests of romantic legend it would be wishfully satisfying if Babbage meant Ada. It seems more likely that he meant either his first love, mathematics, or the Analytical Engine. But he signed the letter unambiguously – 'Farewell my dear and much admired Interpreter'.

Ada's 'Sketch of the Analytical Engine' was a rare accomplishment and confirmed her high regard for her own talents. In it she translated Menabrea's article and added her sometimes imperious notes. Menabrea's article described the general mathematical and operational principles of the Difference and Analytical Engines but gave no mechanical detail. He used the examples of 'programs' that Babbage had taken to Turin to illustrate his lectures. These Ada presented in the form of charts detailing the stepwise sequence of events as the machine progressed through a string of instructions input from punched cards. It is a competent paper and a valuable contemporary document, though it added little new to the understanding of Babbage's schemes. It contained some evocative rhetorical flourishes. 'Who,' asks Menabrea, 'can foresee the consequences of such an invention?' Ada's relish for the metaphysical was probably further inflamed by Menabrea's eulogy to number:

> Yet it is by the laborious route of analysis that [the man of genius] must reach truth; but he cannot pursue this unless guided by numbers; for without numbers it is not given to us to raise the veil which envelopes the mysteries of nature. Thus the idea of constructing an apparatus capable of aiding human weakness in such

researches, is a conception which, being realised, would mark a glorious epoch in the history of the sciences.

She appended to her translation seven 'Translator's Notes' labelled A through G which run to about three times the length of Menabrea's original description. Ostensibly the Notes are expansions of specific points in Menabrea's paper, but they are as much expositions of what she had understood of the machine and its purpose. The examples of 'programs' she used are those worked out by Babbage years earlier, except for the case of Bernoulli numbers which Babbage provided for her afresh. She found a mistake in Babbage's working and sent it back for amendment.

Her taste for mystical speculation was not burdened either by any great mathematical knowledge or by any specific understanding of the mechanics of the machine, leaving her free to speculate about the potential of the Analytical Engine. She provided some fine literary touches. Drawing on the use of punched cards to control the pattern of woven fabrics produced by the Jacquard loom she wrote 'that the Analytical Engine weaves algebraic patterns just as the Jacquard loom weaves flowers and leaves'.

Because of her article 'Sketch of the Analytical Engine', Ada's role in Babbage's work has been both exaggerated and distorted down the years, like a Chinese whisper. She has been described as a mathematical genius, as someone who made an inspirational contribution to the conception and development of the Analytical Engine, as the 'first programmer' and as a prophet of the computer age. It was certainly highly unusual for any woman, especially an aristocratic one, to pursue mathematics with any seriousness. Babbage himself was a highly accomplished and modestly original mathematician with some dozen or so serious mathematical publications to his credit by the time he was thirty. But he discovered no new theorems and left no lasting mathematical legacy. In comparison, Ada was a talented beginner, a precocious novice.

The notion that she made an inspirational contribution to the development of the Engines is not supported by the known chronology of events. The conception and major work on the Analytical Engine were complete before Ada had any contact with the elementary principles of the Engines. The first algorithms or stepwise operations leading to a solution – what we would now recognise as a 'program', though the word was not used by her or by Babbage – were certainly published under her name. But the work had been completed by Babbage much earlier. The charitable view of Ada dates back to her own age. Sophia De Morgan, the wife of Augustus, Ada's mathematics tutor, recalled Ada's response to Babbage's Engine:

> While other visitors gazed at the working of this beautiful instrument with the sort of expression, and I dare say the sort of feeling, that some savages are said to have shown on first seeing a looking-glass or hearing a gun – if, indeed, they had as strong an idea of its marvellousness – Miss Byron, young as she was, understood its working, and saw the great beauty of the invention.

Ada's studies did not cover the elementary mathematics of the Engine (the method of finite differences) until 1841, seven years after her first encounter with the machine. Sophia's account was written fifty years after the event. Posthumous eulogies of this kind, well intentioned as they are, have mythologised both her mathematical abilities and her role in the development of the Engines.

Historians close to the detail of Babbage's work express dismay at the well-intentioned but misguided tributes paid to Ada. Bruce Collier, whose historical study of Babbage's work remains unsurpassed, has this to say about the popular myth of Ada's role:

> There is one subject ancillary to Babbage on which far too *much* has been written, and that is the contributions of Ada Lovelace. It would be only a slight exaggeration to say that Babbage wrote the 'Notes' to Menabrea's paper, but for reasons of his own encouraged the illusion in the minds of Ada and the public that they were authored by her. It is no exaggeration to say that she was a manic depressive with the most amazing delusions about her own talents, and a rather shallow understanding of both Charles Babbage and the Analytical Engine . . . To me, this familiar material [Ada's correspondence with Babbage] seems to make obvious once again that Ada was as mad as a hatter, and contributed little more to the 'Notes' than trouble . . . I will retain an open mind on whether Ada was crazy because of her substance abuse . . . or despite it. I hope nobody feels compelled to write another book on the subject. But, then, I guess *someone* has to be the most overrated figure in the history of computing.

Collier's savagery is perhaps directed less at Ada than at her ill-informed lionisation. Ada has a double allure for historians and compilers of women's biographical dictionaries – daughter of Byron and associate of Babbage. As the daughter of a great poet and the interpreter of the first computer pioneer she seems to reconcile the warring faculties of reason and imagination, of masculine and feminine qualities. She is celebrated as a woman who had apparently defied the oppression of her sex to make a mark in a man's world, and the need for such champions has regrettably distorted her contribution.

Nevertheless, Ada did what none of Babbage's contemporaries did, at least in England: she interpreted and publicised his work on the Analytical Engine. The 'Sketch' was the most substantial account of the Analytical Engine published in English in Babbage's lifetime, and it remains highly revealing of contemporary thinking about automatic computing machines. More specifically, the 'Sketch' includes statements that from a modern perspective appear visionary.

In our search for the roots of the modern computer, we comb contemporary writing for signs of the shift from computers as machines that operate on number to machines capable of manipulating symbols of which number is only one example. During the first two years of design and development, Babbage saw the Analytical Engine as a sophisticated programmable calculator capable of executing any sequence of arithmetic operations, but still operating only on number. In July 1836 he hinted at a more generalised vision in which the engine was seen as a universal algebra machine capable of manipulating not only numbers, but also symbols: 'This day I had for the first time a general but very indistinct conception of the possibility of making an engine work out *algebraic* developments, I mean without *any* reference to the *value* of the letters.' He does not elaborate, though there is an inkling here of a computing engine manipulating symbols according to rules. But the context is still mathematical and it is a tantalising question as to how far he foresaw the power of number to represent entities other than quantity – letters of the alphabet, for example. Ada, however, is explicit and writes that the Analytical Engine

> might act upon other things besides *number* . . .
> Supposing, for instance, that the fundamental relations of pitched sounds in the science of harmony and of musical composition were susceptible of such expression and adaptations, the Engine might compose elaborate and scientific pieces of music of any degree of complexity or extent.

Nowhere, at least in his published writings, does Babbage write in this way nor does he speculate about his machines in other than a mathematical context. Elsewhere Ada writes:

> Many persons . . . imagine that because the business of the Engine is to give its results in *numerical notation* the *nature of its processes* must consequently be *arithmetical* and *numerical*, rather than *algebraical* and *analytical*. This is an error. The engine can arrange and combine its numerical quantities exactly as if they were *letters* or any other *general* symbols; and in fact it might bring out its results in algebraic *notation*, were provisions made accordingly.

As we have seen, Babbage had inklings of a generalised algebra machine, but the ideas are undeveloped and he wrote that 'it would be better to construct a new engine for such purposes'. The Analytical Engine could correctly manipulate + and – signs, and Ada seems to have generalised this capability into a fully fledged algebra. It was not until the 1930s that Alan Turing, the English mathematician and wartime code-breaker, formalised the notion of a computer as a general-purpose symbol manipulator rather than a number cruncher.

Ada echoes a statement by Menabrea that is immediately

recognisable to the artificial intelligentsia: 'The Analytical Engine has no pretensions whatever to *originate* anything. It can do whatever we *know how to order it* to perform.'

It is regrettable that disputes over Ada's 'greatness' continue to mask the study of her ideas and the contribution she made to our understanding of contemporary thinking.

Ada's ambitions for the life of a luminary were tragically thwarted. 'Sketch of the Analytical Engine' was her only publication. Throughout, Babbage remained a close family friend. He ran errands for her in town, escorted her to social events, including the Great Exhibition in 1851, and was a frequent guest at the Lovelaces' homes.

Ada's last months were appallingly grim. She had cancer of the cervix, which remained undiagnosed until late in the day. Her self-righteous and controlling mother Annabella took over the household and carefully controlled access to Ada. She banned Ada's friends, including Babbage, whom Ada attempted to make her executor. In her weakened and pain-ridden state Ada confessed to an adulterous affair with one John Crosse, and to gambling debts. Annabella saw pain as penance and favoured mesmerism, which was fashionable, instead of opium for relief.

The accounts of Ada's last months are heart-rending. If, as Annabella insisted, there is a redemptive value in suffering, Ada surely discharged her sins many times over. Ada died on 27 November 1852 at the age of thirty-six. Babbage's 'dear and much admired interpreter' was buried, on her own instruction, next to her father in the small church at Hucknall Torkard, near Newstead, Nottinghamshire.

Chapter 9

INTRIGUES OF SCIENCE

Every game of skill is susceptible of being played by an automaton.

Charles Babbage, 1864

Domestic life was gloomy, and Babbage seems to have been remote from his sons. The marriage in 1839 of his eldest son Herschel to Laura Jones displeased him, and he was known to disparage Dugald. When Henry, the youngest, left for India as a military cadet in April 1843, Babbage took his farewell in the library, not troubling to see his son to the waiting cab. His indifference was not lost on Henry, then just seventeen. Henry's ship was delayed by bad weather and he was holed up for several days in Portsmouth, where he watched an elderly man entrust his son to the care of the waiting passengers. Henry later recalled that 'he could not help contrasting his tender anxiety for his son with that of my own father'.

While his family life after his wife Georgiana's death was bleak, Babbage's social life was hectic. His diary was crowded with dinner engagements with celebrities, the rich and the titled, frequent stay-overs with the Duke of Somerset, and portrait sittings for the artist Samuel Laurence. Henry recalls that in February 1843 his father 'had no less than thirteen invitations for every day of the month, Sundays included, to dinners and parties of one sort or another'. The social whirl was darkened by the death of Babbage's mother on 5 December 1844. She was in her eighties. Elizabeth had been an unfailing source of moral and emotional support, and Babbage was lovingly devoted to her. The loss must have been a grievous blow.

Suddenly, and for no apparent reason, Babbage stopped the aimless tinkering with the Analytical Engine designs, and in 1846 the Engine work took a new, decisive direction: he began to design a new Difference Engine. The complexity of the Analytical Engine had forced him to ruthlessly reduce the mechanisms to their simplest form, and the economies demanded by the designs suggested simpler and more elegant techniques for the earlier Difference Engine. As Babbage recalled:

> About twenty years after I had commenced the first Difference Engine . . . and after the greater part of these drawings had been completed, I found that almost every contrivance in it had been superseded by new and more simple mechanism, which the Analytical Engine had rendered necessary.

Although his transition from unfocussed pottering to directed design was sudden, there is evidence that he had been actively mindful of the Difference Engine during the development of the Analytical Engine designs. A drawing dating from April 1842 shows an improved layout for a Difference Engine prompted by techniques used in the Analytical Engine. Why he was seized by the design of a new Engine is not clear. It is possible that he had felt a residual obligation to redeem himself and deliver his half of the defunct deal with the Government. It is also possible that the purist in him became seduced by the minimalist elegance with which he could now accomplish something he had previously struggled so much to achieve. Perhaps he simply could not finally relinquish the conviction in the vision only he had glimpsed. It could be that he was just bored.

Between October 1846 and March 1849 Babbage designed his new engine – Difference Engine No. 2. The principle was identical to that of the earlier Difference Engine: calculation by repeated addition according to the method of finite differences. But its execution was a masterpiece of simplicity. The new Engine consists of about 8,000 parts in all, equally split between the calculating section and the printing apparatus. It measures eleven feet long, seven feet high and eighteen inches deep. It has eight columns, each with thirty-one figure wheels arranged in vertical stacks about three feet high. Such was his mastery of mechanism that Difference Engine No. 2 is about three times more efficient than its predecessor – it uses one-third the number of parts for a similar calculating capacity.

For once, Babbage left a complete set of drawings which depict the final design. He had a habit of constant modification – he never could leave well alone. But this set of drawings, twenty-four in all, was spared improvement. As ever, he left no written description of its operation or the rationale of its design.

Difference Engine No. 2 includes a printing apparatus designed to serve both the Analytical and Difference Engines. The apparatus is designed not only to automatically print a hard-copy record of results on a print roll, but also to press the results into soft metal plates or into trays of papier mâché to produce stereotype plates for printing. Astonishingly for its time, it features programmable formatting which allows the user to control the position and appearance of the results on the page. It has options to print down the page in columns with automatic flyback to the top of the next column, or across the page with automatic line-wrap at the end of a line. The margin widths and gaps between columns are alterable, and there is provision for inserting blank lines between groups of lines so that the results appear in blocks for ease of reading. The machine prints in two font sizes simultaneously and can be programmed to print in one, two or three columns.

The printing apparatus plays its part in controlling the engine. As with Difference Engine No. 1, the starting values of the calculation are entered by turning the figure wheels the appropriate amount by hand. Thereafter each turn of the drive handle generates the next value in the table. The moment the drive handle is turned, most of the initial values irreversibly change as the internal state of the machine alters to produce the next result in the table. The results are impressed in turn into the soft material in the printing tray, which automatically advances to receive each new result. If the operator failed to stop in time and tried to calculate more results than the tray can hold, the overrun would spoil the first result on the fresh page and the whole run would have to be redone. To relieve the operator of the anxiety of keeping track of how many cycles the machine has run, and of stopping in time, the printing apparatus automatically halts the Engine dead in its tracks at the end of a page. The mechanism by which it does this is a Heath Robinson contraption which drops a weight into a scoop which operates a lever which pulls a piece of catgut which runs via a series of pulleys to disengage a clutch in the drive. At the appropriate time the handle suddenly runs free, and the Engine is frozen in its tracks ready to resume once the trays receiving the output results are replenished. The printing and stereotyping process has no time overhead at all: the process occurs in parallel with the calculation and does not add to the length of the calculation time. And all this by purely mechanical means.

In 1852 Babbage half-heartedly offered the plans of the new Engine to the new Tory Government under Lord Derby, using Lord Rosse as an intermediary. Rosse was President of the Royal Society, and a friend and supporter of Babbage and his schemes. Babbage was pessimistic about the prospects. Rosse received the reply dated 16 August 1852 from the Prime Minister's office. The outcome was resonantly gloomy:

> Mr. Babbage's projects appear to be so indefinitely expensive, the ultimate success so problematical, and the expenditure certainly so large and so utterly incapable of being calculated, that the Government would not be justified in taking upon itself any further liability.

Babbage had been through too much to protest. He wrote to Rosse saying that he ought to 'make no further attempt to force a generous offer upon a reluctant country' and made a further reference to 'pearls before swine'. He added that he thought the Government was misguided to base its decision on the extravagant costs of the earlier Engine, as the industrial arts had so advanced since then.

Babbage mistakenly thought that the letter from the Prime Minister's office had come from the Chancellor, Benjamin Disraeli, and the rebuff rankled for years. Twelve years on, in 1864, he was still having a dig at Disraeli, calling him a 'novelist and financier'. Disraeli had a celebrated flair for Parliamentary oratory, and Babbage wrote:

> As to any doubt of its mathematical principles, this was excusable in the Chancellor of the Exchequer, who was himself too practically acquainted with the fallibility of his own figures, over which the severe duties of his office had stultified his brilliant imagination. For other figures are dear to him – those of speech, in which it cannot be denied he is indeed pre-eminent.

He grumbles on, scoring elaborate rhetorical points – that the much-abused Difference Engine could 'calculate the millions the ex-Chancellor of the Exchequer squandered' and that 'it may possibly enable him to unmuddle his accounts'.

The set of twenty drawings and four tracings for Difference Engine No. 2 survive unscathed partly because they were never exposed to the hazards of a workshop. They now reside in an archive in the Science Museum Library.

Babbage was an inveterate inventor, and delighted in instruments, contrivances and mechanical novelties of all kinds. Machines were the obsession of the age. Newspapers and scientific journals burst with the wonders of invention, each one more outlandish than the last. His relish for contraptions showed quite early. When about sixteen he devised shoes for walking on water. He attached two hinged flaps – stiff covers from old books – to his boots. The idea was that with the downward thrust of his legs the flaps would spread and the increased resistance to the water would support him. On the upward stroke the flaps would fold back, and by working his legs up and down he could rise up and propel himself. He tried out his miraculous boots in the river Dart in Devonshire. For a short while he succeeded in keeping his head, shoulders and sometimes his arms above water, until one of the flaps got stuck. The hapless lad found himself in the grip of the tide and lopsidedly swam around in circles. Flailing with exertion, he widened the circle to a spiral and eventually made it to the river bank, exhausted.

His near-drowning as a teenager did nothing to dampen his inventiveness in later life. He shed schemes, plans and devices like a postman with a torn postbag. We have already met his plans for cow-catchers, 'black box' recorders for railways, failsafe quick-release couplings for railway carriages, and his calash – the specially built camper he had made in Vienna while on his Continental tours. He designed and constructed countless other devices – a pen with a rotatable disc for drawing broken lines on maps, a chart recorder for logging the condition of railway tracks, and theatre lighting using coloured filters for his 'Rainbow Dance'.

The 'Dance' was a choreographed ballet in which dancers dressed in pure white would take on the hue of the lights projected onto them. The coloured filters were plate-glass cells filled with water in which he dissolved chemicals, and the intense light came from oxy-hydrogen blowlamps. Two fire engines were in attendance for the first dress rehearsal. Dancers representing fireflies flitted in and out of the coloured zones to dramatic effect. But public safety triumphed over art when the theatre management withdrew support on the grounds that the lights were a fire hazard.

There are many other devices he proposed or designed without constructing them – a tugboat for winching vessels upstream, diving bells, a submarine propelled by compressed air, an altimeter for measuring height above sea level, a seismograph for detecting geophysical shocks such as earthquakes, a flat-bottomed boat (a 'hydrofoil') that would aquaplane on water, an astronomical micrometer, a 'coronograph' for producing artificial eclipses, and a lifebuoy with a self-igniting mechanism wound automatically by the bobbing motion of the sea. He was a keen amateur machinist and constructed multi-bladed cutters, several types of tool-holders and an eye-shield to allow close inspection of the cut while machining was in progress.

As a diversion he dabbled in games-playing machines. He came to the conclusion that any game of skill could be played by an automaton. The Analytical Engine already had several of the necessary properties – memory, 'foresight' and the capacity to take alternative courses of action automatically, a feature computer scientists would now call conditional branching. Babbage concluded from this that the key issue of whether or not a machine could play any game, including chess, depended on its ability to store the massive number of combinations of moves. He calculated, perhaps optimistically, that the Analytical Engine was up to the job: 'Allowing one hundred moves on each side for the longest game of chess, I found that the combinations involved in the Analytical Engine enormously surpassed any required, even by the game of chess.' Thus encouraged, he set about designing an automaton for playing noughts and crosses (tick-tack-toe), a piece of cake compared with a chess-playing machine. His investigations threw up fundamental questions of machine behaviour – issues we would now identify as belonging to the study of artificial intelligence. One such question is raised by the issue of 'contention'. If the machine identifies two moves as equally advantageous to winning, what criterion can be used to inform the choice? Since there is no strategic advantage in either option, Babbage let the machine choose arbitrarily. But it is tantalising to see that he used the machine's experience of all past won games as the datum for its choice.

Having dashed off a design for his automaton for playing noughts and crosses, he was struck by the thought of games arcades in which attractively decked-out machines would be challenged by the public. The idea was that the proceeds from these arcades could be used to fund the construction of the Analytical Engine. But he came to the sober conclusion that by the time he had made the machines and set up in business, he would be too old to use the profits. Another wishful scheme of his to fund the Analytical Engine was to write a three-volume novel solely for financial gain. He reckoned that unless he could clear £5,000 it would not be worth his while. He took advice, and abandoned this too.

Yet another of his schemes was his system of delivering mail using an aerial cable-way. In the early 1820s Babbage rigged up a small cable-car system between his front drawing room in Devonshire Street, through the house to the workshops above his stables, and used it to send messages in small canisters to his workmen. He generalised the idea into an intra-city aerial mail-delivery system using wires strung between church steeples, with St Paul's Cathedral acting as a central station from which deliveries could be made every half-hour. He took the scheme even further and speculated about an inter-city funicular mail delivery system with pillars every hundred feet or so and relay stations every three to five miles, offering two or three deliveries a day. It sounds fanciful, but the proposal was apparently meant in all seriousness.

Mechanised communication was clearly a preoccupation. He wrote in some detail on schemes for signalling to and from ships at sea in which a source of light is alternately shuttered or revealed in a coded sequence which contains the message. For one-way communication the occultations of the light are controlled by clockwork, and the preprogrammed sequence of flashes could uniquely identify a lighthouse. For two-way exchanges between ships, or between ship and shore, he developed a communications protocol involving what we now call 'handshaking' which allows the two stations to identify themselves, to 'log on' to each other, and exchange precoded messages. In a description that pre-echoes modern practice, Babbage proposed storing stock questions and answers on removable discs inserted at will into the occulting mechanism for transmission.

He rigged up an experimental signalling device in the upper window of his house facing down Manchester Street – a long straight road – during the Great Exhibition in 1851, when London was teeming with international visitors. The device sent a sequence of flashes coded to represent a number. He took great satisfaction in the fact that observant night-time strollers decoded the number and popped their visiting cards through his door with the occulted number written on it. He observed that it was 'foreigners' who did this – apparently, only visitors from abroad were capable of appreciating his ingenuity.

Always on the lookout to enlist the aid of machinery in the service of human comfort, he installed an elaborate central heating system in his house in Dorset Street in the 1830s. His 'warming and ventilating apparatus' was not the first, but the opportunities for modifying and improving it were endless. He devised and installed a multiple stopcock which allowed him independent control over four separate heating zones, and a self-regulating device to control the furnace. One can picture him having a joyous time monitoring the thermometers, recording their readings and taking great satisfaction in his mastery of nineteenth-century air conditioning.

Some of these contrivances are fanciful, some impractical. Others are not developed beyond speculative proposals. But there is at least one device he invented, built and demonstrated for which he was denied due credit. In 1847 he constructed an ophthalmoscope – an instrument for examining the inside of the eye. He took it to a leading eye specialist, Thomas Wharton-Jones, known to his students at Charing Cross Hospital as 'Mummy Jones'. Wharton-Jones saw no value in the device, and Babbage did not pursue it. Four years later Hermann von Helmholtz was credited with the invention of a similar instrument. In 1854 Wharton-Jones was asked to report on Helmholtz's ophthalmoscope, and he had the grace to own up to his awful blunder:

> Dr. Helmholtz of Königsberg, has the merit of specially inventing the ophthalmoscope. It is but justice that I should here state, that seven years ago Mr. Babbage showed me the model of an instrument for the purpose of looking into the interior of the eye . . . it is much to be regretted for the sake of British ophthalmic surgery, that this discovery was not made public or even utilised at an earlier date.

Wharton-Jones was shortsighted and probably saw only a red blob when he looked into Babbage's instrument. It seems that he was historically myopic as well, and the loser yet again was the luckless Babbage. George Biddell Airy, Babbage's old foe, also had an eye disorder, in his case acute astigmatism. As a student he was plagued by double vision and headaches, and carried three pairs of spectacles with lenses ground to special formulae of his own derived from his knowledge of optics. It seems that those chosen to judge Babbage in his own time were fated to have defective vision.

The ophthalmoscope was not the only inventive feat for which Babbage failed to secure recognition. For many years he had been interested in ciphers – techniques for encoding messages so that only the recipient with the correct 'key' could read them. In 1854 John Thwaites, a dentist from Bristol, claimed to have invented a new cipher. Babbage recognised the technique as the Vigenère Cipher, named after the French diplomat born in 1523, and which, after some 300 years, still defied the best attempts to crack it. Thwaites was peeved that his cipher was not original and challenged Babbage to break it. Babbage found an undetected weakness in the cipher and broke it, a phenomenal achievement which was not publicly disclosed. Nine years later the solution was published by Friedrich Kasiski, who had cracked the cipher independently, and the technique has since been known as the Kasiski Test. It is curious that Babbage did not lay claim to the discovery, if only for the satisfaction of gloating over the vanquished Thwaites. One explanation may lie in the military importance of the solution. Babbage's breakthrough was accomplished not long after the start of the Crimean War, and British intelligence would enjoy a huge advantage over the Russian foe if the technique was still believed to be secure. Whatever the reason for Babbage's silence, the upshot was yet another forfeit of a deserved accolade.

Although Babbage was a prolific inventor, he was strongly opposed to patents. He believed that 'the products of genius' should be freely available to mankind in general and lamented that the poor rewards for scientists obliged them to restrict for profit the wider benefits of invention:

> They may lock up in their own bosoms the mysteries they have penetrated . . . whilst they reap in pecuniary profit the legitimate reward of their exertions. It is open to them, on the other hand, to disclose the secret they have torn from nature, and by allowing mankind to participate with them, to claim at once that splendid reputation which is rarely refused to the

inventors of valuable discoveries in the arts of life. The two courses are rarely compatible.

The case that was exercising Babbage was that of his friend William Hyde Wollaston, who in 1804 had invented a process to produce malleable platinum. Wollaston made no written record of the process and banned visitors from his workshop. He is reported to have made some £30,000 from his monopoly, and kept the process secret until shortly before his death in 1828. Babbage was torn between loyalty to his friend and his utopian principle that science should benefit mankind in an unrestricted way. Wollaston escaped Babbage's mordant censure by revealing the process to a friend during his last days, and the details were published in the *Philosophical Transactions* in 1829 under Wollaston's name. Wollaston was thus redeemed. To Babbage, the eleventh-hour disclosure by Wollaston, 'whose anxiety to render useful even his unfinished speculations, proves that the previous omission was most probably accidental'. Twenty-four years is a long time for an accidental omission, and £30,000 is a substantial sum. Once again, Babbage's friends could do no wrong, and his enemies could do no right.

The Great Exhibition of 1851 was the largest industrial manufacturing extravaganza yet staged. This international festival of the industrial arts was housed in the specially built Crystal Palace – a legendary emporium resembling a massive greenhouse. The construction used about 300,000 glass panes and was built on the south side of Hyde Park. The display halls featured 14,000 exhibitors, domestic and international, and a profusion of products – raw materials, machinery, textiles and fabrics, ceramics and fine arts – organised in thirty sections. The Great Exhibition was a sensation, drawing six million paying visitors in the 141 days for which it was open, between 1 May and 11 October.

This was Babbage's dream world. One imagines him in the thick of it, basking in the advances of his pet subjects and voluble in the advocacy of free trade and the benefits to all of industrial prosperity. The construction of the building used mass-produced frames which allowed speedy construction on site. From his work on the Engines, Babbage, of all people, appreciated the boon to manufacture and construction of using repeated parts. It would seem that he was an ideal candidate to serve on the organising committees or as a Commissioner under Prince Albert, who had earlier won him over with his knowledge of the woven portrait of Jacquard, and who took a direct personal interest in the whole venture. At least he could have expected that the completed portion of Difference Engine No. 1, a celebrated symbol of inventive ingenuity as well as precision engineering, would merit exhibition. Babbage, now sixty, was completely left out. Not just ignored, but actively excluded. His reputation for confrontation and protest as well as his earlier radicalism made him 'unclubbable'.

There had been an earlier snub. He had been bypassed in the lucrative commissions to write one of the eight *Bridgewater Treatises* which were philosophical essays in defence of natural theology. And now, a quarter of a century later, far from being fêted as a proprietor of industrial values, he was simply an outsider. His exclusion was an incomprehensible affront to his self-perception as an elder statesman of the industrial movement, especially as his book *On the Economy of Machinery and Manufactures*, his most successful and acclaimed work, had established him in the 1830s as a leading pundit on the manufacturing arts.

To the *Bridgewater* snub Babbage had responded by offering

in 1827 his own *Ninth Bridgewater Treatise*, uninvited and privately printed, and evidently without embarrassment. He wished to participate in the philosophical debate and his supernumerary essay was a spirited response delivered apparently without rancour. But by the time of the Great Exhibition Babbage had been too badly battered to rise above the injury; instead, he resorted again to public protest. Bitterness infused his pen. In the year of the Great Exhibition he published his vitriolic *The Exposition of 1851; or, Views of the Industry, the Science and the Government of England*. In it he took the Commissioners to task, accusing them of weakness in yielding to retailers' self-interest by omitting the price of manufactured goods on display. He objected to the site chosen for the Crystal Palace, arguing for a site 1300 yards to the east, strategically located for easy access from Cumberland Gate and the gate at Hyde Park Corner. In support of his case he presented calculations to show that with an estimated four million visitors, a total of five million unnecessary miles would have been travelled at a cost of £13,333. Apart from the undertone of protest and indignation, there was nothing outrageous in these arguments. But he did himself no good when he mounted a personal attack on the Astronomer Royal, his old rival George Biddell Airy, who in 1842 had declared his Engines 'worthless'.

In a section called 'Intrigues of Science' Babbage accuses Airy of being part of a vendetta against him, influencing the Government against his Engines through personal jealousy and malice. The villain of the piece is the Reverend Richard Sheepshanks, an astronomer with an early training in law and a close friend of Airy. The circumstances are convoluted and the embroidery is not always easy to unpick. Babbage alleges in *Exposition* that Sir James South, with Babbage's support, twice thwarted Sheepshanks in the politics of scientific affairs.

On the second occasion (a meeting at the Admiralty about the *Nautical Almanac*) Sheepshanks, who relished a scrap, threatened Babbage as they left the meeting room: 'I am determined to put down Sir James South and if you and other respectable men will give him your support, I will put you down.'

The third confrontation between the two men occurred during the notorious court case between South and the instrument-makers Troughton & Simms over an allegedly defective telescope mounting supplied by the company for South's Campden Hill Observatory. The cause célèbre split the scientific community and the 'astronomers' war' was one of the bitterest of the century. The ever combative Sheepshanks was prominent in the hostilities and acted for Edward Troughton. Babbage was lined up on the other side, and testified in 1834 as an 'expert witness' in favour of South. After Babbage had given his testimony he found himself alone with Sheepshanks. Babbage alleged that Sheepshanks again threatened him, saying that his allegiance to South made it necessary to discredit him and that 'he would at some future time, *attack* me publicly on *another subject*' because of his support for South. Babbage took the 'other subject' to mean his calculating Engines.

In each of the confrontations Babbage was alone with Sheepshanks. With Babbage's reputation for indignant outbursts and intemperate public protest there may be the uncharitable suspicion that resentment, anger or paranoia were the author of these episodes rather than verifiable events. Not so. When later challenged Sheepshanks openly admitted the incident, writing that he 'felt great contempt for Mr. Babbage's conduct, and for his mechanical and astronomical ignorance; and I expressed it very openly, and to himself'. Sheepshanks maintained that the 'other subject' was not the calculating Engines but the fact of Babbage's failure to fully discharge his duties as Lucasian professor of mathematics at Cambridge. It seems that, at least in this episode, just because Babbage may have been paranoid, doesn't mean to say he wasn't being got at. In 1839 the court found against South in favour of Troughton & Simms, though this was not the end of it. So bitter was the feud over the telescope that Sheepshanks was pursued beyond the grave. He died in 1856, and the enraged South publicly attacked the contents of the published obituary.

The logic of Babbage's case against Airy reduces to the rather contorted allegation that Sheepshanks was conducting a vendetta against Babbage because of Babbage's support for his friend Sir James South, and that Airy was being personally hostile to Babbage out of sympathy with Sheepshanks, Airy's close friend. The loop was closed with Babbage's imputation that Airy's grudge influenced his judgements against the Engines, and his advice to the Government in opposition to the Engines was correspondingly malicious.

When it came to a shoot-out between Babbage and Airy it was no contest. By 1851, the year Babbage published his attack, Airy's position was unassailable. He had occupied the highest office in civil science for sixteen years, had rendered countless services to the Government, been offered knighthoods in 1835 and 1839 (again in 1853, declining each time but accepting at the fourth time of asking in 1872), and was the most successful consultant engineer in the land. Babbage, at sixty, ten years older than Airy, was seen as a petulant and cantankerous outsider given to volatile outbursts, and the architect of an elaborate experiment to build calculating Engines which failed at spectacular expense to the public purse.

Airy, completely secure, brushed off the attack, disdaining

to make any public defence. Privately he called Babbage's book 'dull' and 'likely to command very little public attention'. He was also amused. Astute and logical, he saw how tenuous was the chain of guilt traced by Babbage to connect a personal grudge of Sheepshanks with his professional advice to the Government via the convoluted route of personal allegiances and the litigation over a telescope. Airy entertained himself by listing eight links in the chain which he enumerated in a letter to one John Barlow, and he ridiculed the drunken progression of Babbage's argument in an eight-verse parodic prose poem along the lines of 'The House that Jack Built'. The third and fourth stanza read as follows:

This is Airy of Greenwich, who holds the Commission
Of Queen's Observator, by Royal permission:
The Government load him with many more cares
*Than the mind of one man with impunity bears**:*
His opinions, in general, are quite undecided:
By Sheepshanks's judgement alone he is guided:
And when Sheepshanks will man or invention assail,
Airy joins in the onslaught with tooth and with nail.
And thus, if a judgement he's called to pronounce,
He is sure in the strongest of terms to denounce

The vilified Engine that Charles built.

* Whether that, which next follows, by this is affected,
Or the two are presented as facts unconnected,
I cannot profess myself perfectly sure,
Mr. Babbage's language is rather obscure.

These are Treasury lords, slightly furnished with sense,
Who the wealth of the nation unfairly dispense:
They know but one man, in the Queen's vast dominion,

Who in things scientific can give an opinion:
And when Babbage for funds for the Engine applied,
They called upon Airy, no doubt, to decide:*
And doubtless adopted, in apathy slavish,
The hostile suggestions of enmity knavish:
The powers of official position abused,
And flatly all further advances refused

For completing the Engine that Charles built.

* Upon this bare assumption, it seems, as a base,
Mr. Babbage has founded the whole of his case;
But I have not remarked, in his book, a citation,
That gives to this notion the least confirmation.

Babbage's grievance against Airy, real or imagined, was sordid linen to wash in public, and the squabble did Babbage no good at all. Friends advised against publication but he paid no heed. The book has been described as 'the diatribe of a disappointed man' and unworthy of him. Meanwhile Airy circulated his ditty and chuckled with his colleagues. Babbage had shot himself in the foot and made himself an object of ridicule.

But *Exposition* is more emotionally revealing than any of his other writings, even his autobiographical work *Passages from the Life of a Philosopher* published thirteen years later. In *Exposition* he wrote pitifully of solitude and loneliness, and revealed the despair to which his efforts, personal sacrifices and lack of recognition had at times reduced him:

> When driven by exhausted means and injured health almost to despair of the achievement of his life's great object – when the brain itself reels beneath the weight its own ambition has imposed, and the world's neglect

> aggravates the throbbings of an overtasked frame, an angel spirit sits beside his couch ministering with gentlest skill to every wish, watching with anxious thought till renovated nature shall admit bolder counsels, then points the way to hope, herself the guardian of his deathless frame.

Self-pitying as this passage is, it is difficult not to be moved. He also paid a deep emotional tribute to his mother, and framed his yearning for recognition as a means of vindicating and rewarding her trust in him:

> He may look with fond and affectionate gratitude on her whose maternal care watched over the dangers of his childhood; who trained his infant mind, and with her own mild power, checking the rash vigour of his youthful days, remained ever the faithful and respected counsellor of his riper age. To gladden the declining years of her who with more than prophetic inspiration, foresaw as woman only can, the distant fame of her beloved offspring, he may well be forgiven the desire for some outward mark of his country's approbation.

Despite despair and depression, he never relinquished the conviction that he had accomplished something of great value, and took solace for the ills of the present from the recognition of posterity:

> The certainty that a future age will repair the injustice of the present, and the knowledge that the more distant the day of reparation, the more he has outstripped the efforts of his contemporaries, may well

> sustain him against the sneers of the ignorant, or the jealousy of rivals.

These passages are startlingly uncharacteristic of any of his other writing which is generally energetic, mischievous and witty. They appear, of all places, in a book supposedly on the Great Exhibition. Babbage was in pain, and his *Exposition*, though criticised as ungracious and unworthy of him, is a rare public testament to the darker moments of his struggles.

Chapter 10

VISIONARIES AND PRAGMATISTS

[Babbage] will then be known for what he truly is – namely, one of the benefactors of mankind, and one among the noblest and most ingenious of the sons of England.

Georg Scheutz, 1857

Georg Scheutz could not believe his luck. He had seen tantalising references to the English calculating engine in the magazines and journals that he avidly scoured for news of technical innovation. Sitting in Stockholm editing, writing and translating into Swedish material for his magazines, he was bursting with frustration at the lack of detail available on Babbage's invention. *Mechanics Magazine*, which he routinely read, had carried a few brief notices and some correspondence. These he translated and embellished with views of his own extolling the virtues of the great invention. But he was starved of the substance of Babbage's schemes and its ambitions. Here, finally, was a godsend – the July issue of *The Edinburgh Review* with a massively lengthy article called 'Babbage's Calculating Engine'. It was 1834. He had been waiting for this for years.

Scheutz, a Stockholm printer, publisher and journalist, was

fascinated by the profusion of new devices pouring out of laboratories and workshops. He was himself an inventor with many patents to his name: an advanced high-speed printing press, chemical dyes, a steam turbine, an optical instrument for copying, a drawing apparatus, and a method of brick-making. Scheutz and Babbage were kindred spirits. Both were prolific inventors and liberal reformers, and committed to free trade and the social benefits of technology. The imagined potential of the calculating engine excited him. As a master of the printing machinery, the Engine's capacity to automatically print and stereotype results inflamed his curiosity.

The article was written by Dionysius Lardner, whose prolific scientific writings and public lectures had brought him wealth, and a degree of celebrity. Sometimes, it must be said, he got it horribly wrong. In 1839 he predicted that high-speed rail travel was impossible because passengers would not be able to breathe. He set the speed above which travellers would asphyxiate at forty miles per hour. He also predicted that crossing the Atlantic by steamship was impossible because of water resistance. Brunel's *Great Western* made the crossing soon after, in April 1838.

Lardner's personal life was equally erratic. While separated from his first wife he eloped with the spouse of a cavalry captain, Richard Heaviside. Husband and father burst in on the runaway couple in a hotel in Paris, forcibly removing Mrs Heaviside and giving the hapless Lardner a sound thrashing. The scandal was reported in the papers on 14 April 1840 and was much tattled over. Heaviside sought damages for seduction and Lardner had to pay up. Adultery was expensive. The shamed Lardner was described in the society gossip of the time as 'the wretched, the deplorably wretched man'.

His colourful personal life and his scientific howlers have made him an object of ridicule. He is described by Anthony

Hyman, historian and biographer of Babbage, as a 'comedy act' and 'a scientific Falstaff . . . even now occasionally mistaken for a serious figure'. But he was a brilliant lecturer and a great disseminator of science. He was one of Babbage's dining circle, and at such gatherings he was doubtless exposed to the promise of the Engine's powers and the saga of its construction. Here was a subject of great public and scientific interest about which virtually nothing had been published. Lardner gorged himself on it.

His article in the *Edinburgh Review* is interminably long. It is bumptious and at times pompous, and glorifies Babbage's attempts to construct his Engines. While preparing the article, Lardner spent time with Babbage in Dorset Street poring over drawings, tinkering with the mechanisms and being instructed by the inventor [on the workings of the device]. In attempting to simplify the principle of the Engine, Lardner used a diagram of thirty-six clock faces to illustrate the sequencing of the calculation, and in doing so confounded the explanation beyond the limits of intelligibility. However, despite the rhetoric, the polemics and the confusing use of clocks, the article gives a comprehensive account of the rationale for the Engine (errors in mathematical tables), the mathematical principle (the method of finite differences) and the state of progress (stalled in a state of stasis for over a year following Clement's walkout in 1833).

For all its detail the article is often obscure, and the rhetoric mystifies the powers of the machine. Babbage had plumbed something fundamental when he observed that there are mathematical sequences that can be computed by his machine according to rules but which cannot be expressed as a general law. In attempting to explain this Lardner makes awed references to arcane powers of the machine. In a passage typical of his rhapsodic pretensions, he writes:

> Yet the very fact of the table being produced by mechanism of an invariable form, and including a distinct principle of mechanical action, renders it quite manifest that *some* general law must exist in every table it produces. But we must dismiss these speculations: we feel it impossible to stretch the powers of our own mind, so as to grasp the probable capabilities of this splendid production of combined mechanical and mathematical genius; much less can we hope to enable others to appreciate them, without being furnished with such means of comprehending them as those with which we have been favoured. Years must in fact elapse, and many enquirers direct their energies to the cultivation of the vast field of research thus opened, before we can fully estimate the extent of this triumph of matter over mind.

Buried in all the puff there is tantalising and suggestive detail elsewhere in the article – that the carriage of tens, for example, is performed as a separate operation after the addition. The article also describes the principle of successive carry and the provision of security devices to prevent derangement of the wheels. But there is no detail at all of the actual mechanisms Babbage used to realise these magical powers.

Georg Scheutz pored over Lardner's account. He was excited both by the metaphysical vagueness of the description and by the possibilities supplied by his own inventive imagination. His excitement was fuelled by his experience of printing machinery – an area in which Babbage had at first been weak. Ignited by Lardner's article, Scheutz, then forty-eight, embarked upon a course that was to secure his ruin: he started to design and build a difference engine. His reasons for doing so changed as he progressed. At first, in 1833, he speculated that just one of Babbage's Engines 'would suffice for the needs of the whole world'. Later, seduced by his own involvement, he hoped to find a market for the Engines in observatories, universities or branches of the printing industry that produced tables. Scheutz saw the Engine, rather than its printed tables, as the saleable product.

Without details of Babbage's mechanisms, Scheutz devised his own. While his mechanisms performed the same logical function as Babbage's they were completely original. His schemes remained on paper until 1837 when Scheutz's son, Edvard, then a lad of fifteen, offered to build a working difference engine using his practical training in engineering drawing and craft skills – metal-turning, woodworking, and foundry work. He developed his father's ideas by trial and error and by the end of summer 1843 had succeeded in building a working machine. The device was modest in that it operated with only three orders of difference, but it included a working printing mechanism – an essential feature of Babbage's scheme – which stereotyped results to seven figures. In six years Edvard, with the encouragement of his father, had built the first automatic printing calculator.

Babbage had demanded the highest precision in the manufacture of parts for his Engine, and British technology, which was the most advanced anywhere, was stretched to its limits and beyond. Edvard's machine has a rough wooden frame and was made using a simple lathe and hand tools by a young man with craft skills. A Swedish teenager had succeeded where the best of British had failed. The successful completion of the Swedish prototype raises questions about whether Babbage's demand for the highest precision was warranted by the needs of the mechanisms.

With the prototype under his belt, Georg Scheutz got cracking with the task of promoting the invention. He offered a fully engineered machine in metal to the British Government, via the Swedish Embassy in London. An advisory board was convened and Airy, Babbage's *bête noir*, was invited to officiate. There is no official record of their findings, though Georg records that the British authorities turned him down on the grounds that Parliament would not support a foreign invention of the same kind as the English Difference Engine which had already proved so costly. Airy knew long before Babbage that there was a new engine being built in Sweden.

The failure to interest the authorities in England was followed by similar failures to secure state support in Sweden. Finally, in 1851 Scheutz senior made an appeal to King Oscar for the modest sum of £283. It seems that Babbage was not alone when it came to appealing to the mighty over the heads of bureaucrats. The sum was eventually granted, though there were binding conditions on its award: payments would be made in instalments and be subject to the assurance from the Swedish Academy of Sciences of financial propriety and progress; funds would be paid only against personal security signed by Scheutz himself; and there was a fixed deadline for completion which if not met would render Scheutz personally liable to refund all moneys received.

The contrast between these arrangements to safeguard a relatively trivial capital sum and the offhand deal between Babbage and the British Treasury is stark. The Treasury's commitment was vague and the funds were vast. There was no deadline, and payments were not conditional on progress, completion or the success of the physical result. The Swedish machine was completed in October 1853 – ahead of the deadline. The handsome engine, an expanded version of the earlier prototype, was professionally engineered in metal. It calculated with four orders of difference to fifteen decimal places and automatically printed results to eight digits. The Scheutz machine was smaller than Babbage's Difference Engine. It was more delicate, just as complex and about four times slower. But it had one great virtue – it was finished.

The Swedish engine was brought to London in search of a buyer. It was widely known on the Continent that Babbage's Engine project was long since defunct, and Whitehall's generosity to Babbage was seen as official acceptance of the value of calculating engines. The Scheutzes saw England as a mature market. The machine was installed at the Royal Society, Somerset House, in November 1854.

Like its predecessors it became an object of learned curiosity for dignitaries of the scientific establishment to cluck over. Luminaries of the day as well as the officers and Fellows of the Society were among those who viewed the new wonder – Peter Mark Roget (physician, savant and originator of the eponymous Thesaurus), Charles Wheatstone (pioneer of the telegraph and inventor of the concertina) and Michael Faraday (famous electrician) were among the curious. Prince Albert visited in June 1855. The engine was royally inspected and the event was big news. The *Illustrated London News* covered the event with a near-full-page spread featuring a large engraving of the engine and gushing description of its wonders, one of which was that it needed so little physical force to operate that it could be driven by a small dog used to turn roasting spits.

The Scheutzes were worried about how Babbage would react to their success. They feared that the great man might see them as competitors, or even as intellectual plagiarists. To the Scheutzes' great relief, Babbage welcomed them with open arms. He showed them his house, his workshop and the partially completed engine, and spent two days exploring the Scheutzes' machine. Babbage had by this time given up on his own Difference Engine and was adamant that he had 'no intention under any circumstances of ever making such a machine' himself. Of all people, Babbage was someone who could empathise with the political and financial tribulations that marked the twenty years it had taken the Swedes to realise their machine. Any misgivings the Swedes may have had were banished. Far from competing with the Scheutzes, Babbage devoted himself to promoting their interests.

Airy should have been the Scheutzes' most prized potential customer. As Astronomer Royal, as director of the Royal Observatory at Greenwich, which was the envy of its European counterparts, and as chief scientific adviser to the Government, Airy must have topped the client list. But he is curiously absent. He was not specifically invited to view the machine, and he was not included in the four-man committee appointed by the Royal Society to report on the new engine. It is not difficult to imagine that Babbage was behind this exclusion. There was now a warm allegiance between Babbage and the Swedish visitors, and as intellectual host and strategic adviser to the Scheutzes Babbage would surely have alerted his new friends to Airy's antagonism to calculating engines and taken early measures to keep the forthright Royal Astronomer well out of the picture.

But Airy would not be silenced. He contrived to view the engine privately and have it demonstrated, and when a transcript of the Royal Society report on the engine was published in the *Philosophical Magazine* in July 1856, he wrote to the editors. The Society's assessment of the machine was soberly favourable, though not ecstatic. Airy wished to dampen even these modest endorsements:

> I cannot refrain from expressing my general admiration of the beauty of its arrangements and my assurance that . . . it can be constructed at small expense. I am

> also impressed with the ability and accuracy of the Committee's Report. But I wish to guard myself from giving an opinion on the utility of the machine; remarking only that, as I believe, the demand for such machines has arisen on the side, not of computers, but of mechanists.

His allegation is clear: that the motive for calculating engines sprang not from computational need but from the passions or entrepreneurial drive of technologists. Certainly neither Babbage nor the Scheutzes were too clear about the practical utility of their machines. Rarely if ever was their advocacy graced by anything resembling a calculation or an argument for cost effectiveness. Babbage was always rather vague about actual benefits, and he left it to the likes of Lardner to bang on about the damnation of tabular errors and the glorious enterprise of science. The Scheutzes had a more obvious commercial interest in the engines, but they too were curiously casual about market research.

The Astronomer Royal's letter to the editor went on to list the four stages used to prepare typical tables, and analyse how each stage might or might not benefit by mechanisation. Airy's arguments finally reveal the technical grounds for his consistent scepticism of the utility of calculating engines, and his views are profoundly damaging to the utopian ideal of a handle-cranking solution to the error-free production of mathematical tables so dear to the advocates of the engines. He was a pragmatist whose criterion of usefulness was based almost entirely on the direct benefits to existing tabulation practices at the Royal Observatory, which he ran. Babbage, and to some extent the Scheutzes, were visionaries, intoxicated by ingenuity, intricacy, the mastery of mechanism and the seductive appeal of control over number.

Airy wrote with extraordinary clearsightedness. He suggested for the first time that the real value of difference engines is not to generate new tables from scratch using repeated addition, but to detect errors in existing tables through repeated subtraction. Babbage seemed to have missed this trick. Indeed, when commercial machines capable of differencing finally became available in the 1920s and 1930s they were used for error-detection in existing tables rather than for the generation of new ones. Airy's prescience here is remarkable. He showed through his letter that his condemnation of the engines was better founded than Babbage ever gave him credit for.

The Scheutz difference engine was exhibited in Paris in 1855 at the Great Exposition, where it won a gold medal. Babbage's son Henry, in England on furlough from India, had carefully prepared the symbolic description for the Swedish machine using his father's favoured Mechanical Notation. Babbage and two of his workmen dashed over to Paris with the reams of calico on which the notations had been pasted. But the impulsive Babbage had no official status, and he was refused admission.

After the Exposition the engine was transferred on the instruction of Napoleon III to the Imperial Observatory in Paris, where it was examined by the astronomer Urbain Le Verrier with a view to purchase. The asking price was 50,000 francs (then about £2,000). Le Verrier could see little use for the machine, and the calculations that might be of use 'it did less quickly than an ordinary calculator'. The purchase was rejected. The engine returned to London, where it was used to produce a promotional booklet, *Specimen Tables Calculated and Stereomoulded by the Swedish Calculating Machine*. *Specimens* was dedicated to Babbage 'by his sincere admirers Georg and Edward Scheutz'. The Preface carried a glowing tribute to their English host, claiming that when the usefulness of the calculating engine was finally proven Babbage 'will then be known for what he truly is – namely, one of the benefactors of mankind, and one among the noblest and most ingenious of the sons of England'.

The fifty-page brochure was sent to everyone with an interest in tables that Babbage and the Scheutzes could think of. The circulation list runs to about 400 names, with world-wide distribution to observatories, academies, universities, libraries, banks, newspapers, astronomers, civil servants, presidents, emperors, kings, queens and princes. The Swedish engine eventually fulfilled the destiny so fervently desired by its architects. It was purchased – by an American. With Babbage as broker, the engine sold for £1,000 to the Dudley Observatory in Albany, New York, a flagship project with ambitions to become the 'American Greenwich'. The machine arrived in Albany in April 1857, where it languished unused. It is now a museum piece in the Smithsonian Institution in Washington.

The Scheutzes' efforts to market their engine had caught the attention of the chief statistician at the General Register Office, William Farr, who for some time had been watching the unfolding drama. Farr enjoyed a formidable reputation as England's foremost medical statistician. He was an imposing figure, forthright and dedicated, and such was his influence that he was often mistaken for the Registrar General. The General Register Office was responsible for analysing official records of births and deaths, as well as the mass of data from the national census held every ten years. Recording, compiling, calculating and tabulating were the chief statistician's everyday fare, and by the mid-1850s Farr had vast experience of preparing tables of life expectancy, annuities, insurance premiums, and interest payments, which were published in the mammoth volumes of the *English Life Table*. He also had years of experience of collating, analysing and tabulating data from the censuses. The advantages of mechanised calculation to relieve the numbing tedium of manual methods were evident to him early on.

Farr's background was impoverished. After a medical training in provincial Shrewsbury and then in Paris, he returned to practise in England. The medical profession was class-ridden, hierarchical and competitive. Without wealth, connections or gentlemanly graces Farr found himself disadvantaged, and he turned to medical journalism and statistics. He finally abandoned medicine to join the General Register Office in 1839, where he served as a career civil servant for forty years, committed to the use of statistics as an instrument of social reform and tireless in his efforts to alleviate the plight of the poor through improved public health.

The belief in statistics as salvation from suffering is reflected in a fictionalised account that appears in Peter Ackroyd's novel *Dan Leno and the Limehouse Golem*, published in 1994. In this account Babbage completed his Analytical Engine, which is housed in a factory on Commercial Road in the impoverished East End of late-nineteenth-century London. In the fictionalised account Babbage had died nine years earlier, and the machine's superintendent is Turner, his foreman of twenty years. To a visitor who suggests that corruption is a result of poverty, Turner replies:

> That was going to be my own precise point, sir. That was what Mr. Babbage also believed: if only we could calculate the incidence and growth of the poor, then we could take proper measures to alleviate their condition. I have lived along Commercial Road for many years, sir, and I know that with notation and data we might take away all the sorrow.

While the Scheutzes' engine was on exhibition at the Royal Society Farr tested it and impressed some results into stereotype moulds, and was well pleased with the machine's performance. He had plans for an ambitious expansion of the *English Life Table* and saw the engine as an affordable alternative to hiring squads of additional clerks and computers.

In the summer of 1857, Scheutz, Farr and his boss, the Registrar General, George Graham, launched a three-pronged attack on the Chancellor, George Lewis, to secure the sum of £1,200 from public funds for a copy of the Scheutz engine. To avoid outright rejection, Graham suggested to the Chancellor that the engine might be useful not only to the General Register Office but to others – the Astronomer Royal, for example, or the Nautical Almanac Office. If this was a ploy to force wider consultation, it worked. Airy, chief scientific adviser to the Government, was asked for his opinion. His role was twofold. As the director of the Royal Observatory at Greenwich he would comment on its potential use there, and as an adviser to the Treasury he would counsel on its use at the Nautical Almanac Office and at the General Register Office and make a general recommendation.

The last occasion on which Airy had been consulted by the Treasury was fifteen years earlier, when the Chancellor, Henry Goulburn, had asked for his confidential opinion on the wisdom of further government investment in Babbage's Engine. Then, Airy had savaged Babbage and his hopes. This time, he had an official brief to comment on the possible uses of the machine in three areas of government work, and his personal views could not properly feature in any recommendation. Airy was a conscientious civil servant, and he compiled a questionnaire which he sent to the Registrar General, George Graham, and the superintendent of the Nautical Almanac Office, J. R. Hind. The questions were highly specific. How often were new tables constructed using the method of differences? Was a permanent record of results required? Was a record of intermediate steps in the calculation desirable? If a machine was on hand, would it be preferable to use a man to run it, or compute with pen and paper? And so on.

The response from the Nautical Almanac Office was uniformly negative. The response from the General Register Office was determinedly positive and backed up by an authoritative case from Farr. Farr's credible and technically informed case flew in the face of Airy's own long-held scepticism about the usefulness of the engines. In September 1857, Airy submitted his report. He wrote:

> In the Royal Observatory, the Machine would be entirely useless . . . During the twenty-two years in which I have been connected with the Royal Observatory, not a single instance has occurred in which there was a need of such calculations.

His conclusions on the use of the machine by the Nautical Almanac Office are more considered, but equally dismissive:

> [The superintendent's] opinion is clear and unhesitating, that no advantage would be gained by the use of the Machine, and that he would prefer the pen-computation of human computers, in the way in which it has hitherto been employed.

But Farr was not so easy to dismiss, and it is possible that the quality of Farr's advocacy had seeded some genuine curiosity about the utility of the machine. Despite his misgivings, Airy finally recommended that a copy of the Scheutz engine be purchased from public funds for trials in the General Register Office. Farr had effected the first reluctant softening in the Astronomer Royal's steadfast opposition to the machines.

The machine was built at a thumping loss by Donkin & Co., and completed in July 1859 several weeks past the deadline. Costs overran and Bryan Donkin was out of pocket by over £600. The engine was used as Farr had intended – for the production of the much expanded *English Life Table* published in 1864. But the hoped-for benefits of mechanised tabulation were nowhere realised. The machine included none of Babbage's security mechanisms to safeguard against derangement. It needed incessant attention, and Farr was in a state of constant anxiety about whether it had strayed. They ran it at night. Farr told of how 'its work had to be watched with anxiety, and its arithmetical music had to be elicited by frequent tuning and skilful handling, in the quiet most congenial to such productions.' It was evidently a delicate flower. The utopian ideal of guaranteed error-free results at the casual turn of a handle was a long way off.

In the event, the machine made only a slight contribution to the 1864 *English Life Table*. Of the 600 pages of printed tables in the volume, only twenty-eight pages were composed entirely by the machine. In all, 216 pages were partially composed and the rest were typeset by hand. This was not all. The hoped-for economies from automatic stereotyping also evaporated. It was expected that if the machine produced stereotype plates for printing, the costs of typesetting and checking would be saved. Not so. Her Majesty's Stationery Office, which produced the fat volume, stated baldly that had the entire volume been automatically typeset by the machine it would have made a saving of only ten per cent over conventional methods. Despite Farr's hopes and his valiant campaign, the Scheutz engine failed to deliver any significant technical or financial benefit. It seems that Airy, so vilified by Babbage for his hostility to the engines, had been right all along.

But there is one last twist to the tale. Airy's implacable opposition to the engines started to crack. When Donkin delivered the Scheutz engine to the General Register Office, George Graham asked Airy to perform an acceptance inspection and sign the machine off so that Donkin could be paid. This was not a stipulation of the original arrangement, but Graham evidently felt the need for expert confirmation before settling Donkin's bill. Airy was in no rush, and it took Farr seven weeks to pin him down.

Airy finally put the machine through its paces on 30 August 1859. His report, dated the following day, commended the construction, and he pronounced himself entirely satisfied. But something happened to Airy during the close examination of the machine. After decades of dismissal he suddenly saw a use: producing tables suitable for astronomical use which required results in degrees, minutes and seconds of arc rather than the standard decimal format used by both Babbage and Scheutz. He wrote immediately to Donkin asking if the machine could be adapted for such use in the production of the *Nautical Almanac*, and whether the existing machine could be used interchangeably for both purposes. He also asked what a new machine would cost if the two uses were incompatible.

He was clearly much taken with the new possibilities, and without waiting for a reply from Donkin wrote the next day to William Gravatt, an engineer who had been associated with both Babbage's and Scheutz's engines. He asked again about the implications of conversion from decimal to sexagesimal, and urged Gravatt to consider his earlier suggestion that the real value of difference engines is not to generate new tables from scratch using repeated addition, but to detect errors in existing tables or new manually computed tables using repeated subtraction. Airy may not have become a card-carrying zealot but he had certainly undergone a partial conversion. Donkin replied that the Scheutz engine was already capable of sexagesimal operation, and that the conversion was trivial in time and cost.

It is difficult to know whether Airy's sudden softening was a transient lapse, or whether he might actually have melted had the Scheutz engine been less in demand in the months that followed. Either way, after this flurry of activity Airy bowed out and took no further known part in promoting the use of the engine in the Nautical Almanac Office, or anywhere else. If Babbage was aware of Airy's volte-face, there is no evidence of any reaction from him.

Georg Scheutz died a bankrupt in 1873 at the age of eighty-seven. Edvard died bankrupt in 1881 aged fifty-nine. Their crusade to build and market calculating engines contributed to their ruin. The Scheutz-Donkin difference engine is now in the Science Museum in London.

The movement to automate calculation had failed. There was no applause for the heroic pioneers. Just a long silence.

Chapter 11

CURTAIN CALL

It's a long time coming.

Charles Babbage, 1871

In June 1857 Babbage dusted off the drawings and restarted work on the Analytical Engine after a ten-year gap. There were no more inspirational spurts and no great breakthroughs. But the new phase of work was sharply directed towards finally constructing a machine, and he started resolving issues of the design that he had left open. He changed the configuration of the Engine to make it more compact and designed complex machine tools for its construction. He had dithered about the Engine's capacity. Now he fixed on a machine with 1,000 columns in the Store, each with fifty digits, though he changed his mind several times after. He corresponded at length with Whitworth in Manchester, placing orders for parts and tools.

Time was running out. The old master was approaching seventy, and his faculties were failing. An acquaintance who first met Babbage in autumn 1861 recalled that 'though he still retained much power of thought, he had lost the faculty of arranging his ideas, and of recalling them at will'.

In March 1859 Babbage estimated that he could complete the Analytical Engine in two or at most three years. Six months later work stopped again and was not resumed for over a year. In 1863 he wrote that given five or six years more he may yet see the machine at work. He had revised his estimate but had not yet given up.

Interleaved between bouts of work on the Engine he found time to write his rambling and entertaining reminiscences *Passages from the Life of a Philosopher*, which appeared in 1864 – the same year as Farr's *English Life Table*. *Passages* is a lively collection of yarns. Despite its overall cheeriness he could not resist raking over old quarrels and settling old scores. Touchy and proud, he never forgot a slight or an injury, and, true to form, he ruffled a good few feathers with his accounts of real and imagined transgressions by his erstwhile colleagues and associates.

The publication of *Passages* lifted his reputation for public protest to new heights. During the 1850s he was driven to distraction by street musicians, and *Passages* includes a bitter attack on the curse of 'vile and discordant music'. When he first moved to Dorset Street in 1829 the neighbourhood was quiet and genteel. As metropolitan London grew, the bustle spread. A hackney-carriage stand appeared in his street, and respectable tradesmen with whom Babbage had dealt for twenty-five years moved out. The area deteriorated as coffee shops, pubs and lodging houses opened up. And with new transient clientele came the organ-grinders and buskers who relentlessly cranked out a quasi-musical din. While the handles on the barrel organs whirled round, those on his own machines were quite still.

The cacophony from bands, organs, trumpets, hurdy-gurdies and drums drove Babbage to despair. He campaigned against the noise-makers for all he was worth. He had them arraigned by the police, sought prosecutions and fines through the local magistrates, and bribed the tuneless scourges to go away. The street artistes had a vested interest in being abominably bad – it would hasten the bribe and increase its size. It could cost a shilling to a half-crown to buy them off. They would then often just carry on.

With habitual precision, Babbage recorded that in an eighty-day period he had suffered 165 disturbances requiring his personal intervention. The effect on his work and concentration was devastating. With helpless fury he calculated that over a twelve-year period 'one-fourth part of my working power has been destroyed by the nuisance against which I have protested'. Dickens's Gradgrind would have applauded such quantification of despair. But his fact-gathering scarcely conceals the frustration at his powerlessness to stop the blight. He took to taking walks during the day to avoid the din, and to working at night well into the small hours. But even the sanctuary of darkness was invaded. A sheet of his notebook is savaged by the stab of a pen dragged down in an ugly tear when, hounded late into the night, he could take it no longer.

Babbage was not alone in his campaign against street nuisances. It was a recognised problem of metropolitan life, and when *Passages* appeared there were already moves afoot at Westminster for legislation. A bill was passed in 1864 and Babbage's crusade had helped to raise public awareness. But the vehemence of his attacks on the perpetrators made him a figure of fun. He was followed and taunted by throngs of street urchins as he agitatedly went in search of a policeman to arrest an offender. He was pursued by marching bands as he went about his business, baited and provoked by neighbours who hired out-of-tune players to concert outside his house. As if this was not enough, he received anonymous letters threatening his house and person.

In his campaign against the musicians he repeatedly invoked the damage of public disturbances to those ill or convalescing. Appealing as ever to the authority of number, he estimated that since 4.72% of the inhabitants of London were ill at any time (the figure is derived from benefit society returns) there would be on average twenty-six people ill in his street alone whose recovery would be impaired by the 'abominable nuisances'. The argument has particular poignancy, as we shall see.

His hypersensitivity to the issue of street musicians sabotaged a rare occasion on which Babbage demonstrated the Difference Engine in public. The finished portion of the unfinished Difference Engine No. 1 was exhibited at the Exhibition of 1862 in London. Babbage was aggrieved at the small display area allocated by the Commissioners. 'English Engine poked into hole,' he growled. Despite this the Engine attracted large crowds, and the engineer William Gravatt conducted demonstrations. Occasionally Babbage took groups of friends to see the machine and explain its wonders to them. Crowds tended to gather, and on one occasion Babbage was loudly heckled from the floor by someone questioning his taste in music, especially organ music. Babbage ignored the incident, but a few days later he was again barracked and ridiculed. He quit his demonstrations in a huff, and left the task to Gravatt, rarely going near the Engine again.

His assault in *Passages* on the curse of street nuisances is often regarded as evidence of eccentricity, and it is tempting to see Babbage, even affectionately, as a comic figure astride one of his frisky hobby horses. But before we snigger with his tormentors, there is a new factor that we should take into account. Babbage's autopsy report was found amongst family papers in 1983 by his great-great-grandson, Neville Babbage, a medical practitioner in Australia. The post-mortem examination was carried out seventeen hours after death. The weather was warm and damp. The examination found a form of arterial disease that is now known to cause degeneration of the inner ear resulting in a hearing disorder. Babbage's acute sensitivity to noise and his intolerance of discordant disturbance may well have had a medical origin, and history should perhaps modify its ridicule to something less harsh.

Yet still he tinkered with the Analytical Engine. In the late 1860s he experimented with methods of making the repeated parts he needed. He pioneered pressure-diecasting techniques in which molten metal was forced into moulds hollowed out to the shape of the part, and explored methods for stamping components from sheet metal using punches and dies. He completed perhaps half the parts for a small simplified experimental model of the Mill of the Analytical Engine. But he ran out of time. The task was huge, his energies were waning and he never could settle on a final version. An anecdotal portrait of Babbage's disheartening complacency about perpetual incompletion is provided by Lord Moulton in an inaugural address delivered in 1914. When Moulton visited Babbage, the elderly inventor took him on a tour of his workshop. Moulton recalled:

> In the first room I saw the parts of the original Calculating Machine, which had been shown in an incomplete state many years before and had even been put to some use. I asked him about its present form. 'I have not finished it because in working at it I came on the idea of my Analytical Machine, which would do all that it was capable of doing and much more.

> Indeed, the idea was so much simpler that it would have taken more work to complete the Calculating Machine than to design and construct the other in its entirety, so I turned my attention to the Analytical Machine.' After a few minutes' talk we went into the next work-room, where he showed and explained to me the working of the elements of the Analytical Machine. I asked if I could see it. 'I have never completed it,' he said, 'because I hit upon an idea of doing the same thing by a different and far more effective method, and this rendered it useless to proceed on the old lines.' Then we went into the third room. There lay scattered bits of mechanism but I saw no trace of any working machine. Very cautiously I approached the subject, and received the dreaded answer, 'It is not constructed yet, but I am working at it, and it will take less time to construct it altogether than it would have taken to complete the Analytical Machine from the stage in which I left it.' I took leave of the old man with a heavy heart.

In 1868 Babbage's last publication appeared – an article on geology. In all he had published eighty-six scientific and miscellaneous papers, and six books. The scope of his work was broad even by the generous standards of Victorian polymathy – mathematics, chess, lock-picking, taxation, life assurance, geology, politics, philosophy, electricity and magnetism, instrumentation, statistics, railways, machine tools, political economics, diving apparatus, submarines, navigation, travel, philology, cryptanalysis, industrial arts and manufacture, astronomy, lighthouses, ordnance and archaeology.

Babbage's last years were wretched. He was plagued by

nightmares and hallucinations. He woke frequently, distressed, fatigued and often with severe headaches. He was also bitter and resentful at the dismal outcome of his labour. A friend who saw him towards the end commented that 'he spoke as if he hated mankind in general, Englishmen in particular, and the English Government most of all'.

In March 1871 Babbage's youngest son, Henry, accompanied by his wife Min, arrived back in England on furlough from military service in India. Henry spent more time at Dorset Street as his father's health deteriorated, and was surely some comfort when Babbage learned of John Herschel's death in May that year. In a letter of condolence to Lady Herschel Babbage referred to Herschel as 'one of the earliest and most valued friends of my life'.

On Thursday 5 October Henry received a telegram from Richard Wright, Babbage's workman, that the old master was ill. Henry travelled immediately to London from Bromley and found his father in great pain. He took charge of household affairs, secured a trained nurse, and anxiously awaited bulletins from Sir James Paget, the attending physician.

The ailing Babbage started having shivering fits. On Sunday the 8th his rest was disturbed by a persistent organ-grinder nearby. Henry had already complained to the Commissioner of Police, but without result. Min joined Henry at Dorset Street. Babbage changed the will he had drafted in 1849. The main alteration was to bequeath the Engines, parts and drawings to Henry, who had involved himself in the work during a long period of leave in the mid-1850s. The will was signed on Friday the 13th.

Babbage knew he was dying. Weak and shivering, he said to Henry, 'It's a long time coming.' In a relentless twist his final hours were plagued by organ-grinders seemingly determined to hound him to his grave. Henry recalled the last moments:

> On the 16th the organ-grinders were particularly troublesome, and before I went to sleep I wrote again to the Commissioner of Police, but the organs were playing all the same on 17th, both about midday and about 9 p.m, and in the afternoon there was a man inciting boys to make a row with an old tin pail . . . but no policeman in sight. At 8.45 p.m. there was again more organ playing. On 18th organs playing again about midday . . . C.B. passed away about thirty-five minutes past 11 o'clock p.m. Shortly before he had said, 'What o'clock is it, Henry?' and I had told him.

Charles Babbage died on Wednesday 18 October 1871, just over two months before his eightieth birthday. He was buried in Kensal Green cemetery six days later. There was only one carriage – the Duchess of Somerset's – and few mourners.

Part III

A Modern Sequel

Chapter 12

HERE WE GO AGAIN

I have no intention under any circumstances of ever making such a machine myself.

Charles Babbage, 1855

I was at my desk in a gloomy office on the first floor of the old Meteorology Block of the Science Museum in South Kensington, poring over the inventory lists of the collections. It was a bright spring day in 1985, though the office, which had only a small north-facing window, was largely dark. Pale light, dappled by the foliage of the trees between the Museum building and neighbouring Imperial College, fell on scattered papers in front of me. I was working through the long lists of objects in the collections I had recently inherited as the new curator of computing, trying to imagine the objects described by the short, sometimes cryptic, inventory entries. I became aware of a movement to my right. Marking my place on the page with my finger, I looked up. A smiling bearded man wearing a waistcoat stood at my door. He held in his hand three sheets of paper.

My visitor introduced himself as Dr Allan Bromley, from the Basser Department of Computer Science, Sydney, Australia. I invited him in. He sat down and I listened to his tale. Bromley explained that he had been studying Babbage's design drawings in the Science Museum Library during a series of visits going back to 1979. He had developed a detailed understanding of the Engines and was convinced that at least one of them could be built. The document he carried, fresh from the typewriter, was a proposal to build a Babbage Engine. The proposal was dated 20 May 1985 and addressed to the then Director of the Science Museum, Dame Margaret Weston, who had encouraged him in his studies of Babbage's work.

The timing of Bromley's arrival was uncanny. I had taken up my appointment at the Science Museum just a few weeks before, and had been reading about Babbage in the books left behind by my predecessor. What astonished me was that the tale of Babbage's efforts remained unfinished. I found it incomprehensible that no one had attempted to build a Babbage Engine once the accepted reason for his failure (the limitations of mechanical engineering in the nineteenth century) no longer applied. How could history leave so compelling a question unanswered? Why had no one been intrigued enough to build one of his machines?

I later learned that I was not the first to wonder why. The historian Anthony Hyman had fallen under Babbage's spell while researching his biography *Charles Babbage: Pioneer of the Computer*, published in 1982. Hyman spotted that the Manpower Services Commission had sponsored the construction of a replica of Stephenson's locomotive, the 'Rocket', which won the Rainhill Trials in 1829. Hyman wrote to *The Times*, and in a piece headed CALCULATING OUR DEBT proposed that the Commission should fund the construction of working versions of Babbage's calculating Engines, to commemorate Information Technology Year. Taking a leaf out of Babbage's book, he appealed to government. He sent the proposal to Kenneth Baker, Minister of State for Industry. The response was a polite rejection. Hyman, an energetic champion of Babbage's work, was convinced that the Engines could be built, but without financial resources and a detailed technical analysis of the engine designs, his initiative, like others before it, came to naught.

Bromley's proposal was concise and clear:

> My studies of the design of Difference Engine No. 2 convince me that it would have worked. The logical or mathematical design is simple and elegant. The mechanical design had been refined by Babbage's many years of design experience and incorporates many neat devices to ensure that the mechanism could not, by accident, become deranged and the calculation fall into error. The calculating, printing and control mechanisms all exhibit a similar elegance of design. All parts should function correctly, separately and in unison, if made to the accuracy known to have been attained for the parts of the earlier Difference Engine.

Bromley was better placed than anyone to assess the feasibility of an engine construction. He had conducted the most detailed analysis of the surviving drawings and decoded some of the most obscure and complex features of the designs.

Babbage's reputation as a computer pioneer rests more on the Analytical Engine, a programmable general-purpose machine, than on the better-known Difference Engines. Yet Bromley proposed, as had Hyman before him, that we build not an Analytical Engine but a Difference Engine. There are practical as well as historical reasons for preferring the earlier machine. Attempting an Analytical Engine is a mammoth task. Babbage constantly changed his designs by adding new improvements. Drawings had the perpetual status of 'work in progress', and at any given time there were aspects of the engines that were incompletely specified. It is also misleading to refer to the Analytical Engine as though it is a single defined entity. There was a series of Analytical Engines of different sizes and capacities. Before building an Analytical Engine, a generational version would have to be chosen, and then a definitive design completed – formidable obstacles to be overcome before construction could start. There is also the issue of credibility. Since Babbage had never himself completed any of his Engines, there remained the question of whether his designs were elaborate but impractical dreams, or realistic engineering specifications. An Analytical Engine was simply too ambitious a project to undertake, both technically and financially, without some simpler staging post.

There was another feature of Babbage's legacy that lent appeal to Difference Engine No. 2: there is no evidence that any of the drawings for it are missing. Although the designs were half-heartedly offered to the Government in 1852, no attempt had been made to construct the Engine, and no part of the Engine had been attempted since. The drawings had never been split up for the execution of different tasks, and never exposed to the hazards of dirt and damage in a workshop. The same cannot be said for the drawings for Difference Engine No. 1, which were both well used and well travelled. They were transported to Clement's workshop, returned to Babbage, migrated to King's College, then to the old South Kensington Museum, and ended up in the Science Museum archive, passing through the Victoria & Albert Museum en route. Some of the drawings are missing and others did not survive unscathed. Not only does the set of twenty design drawings and several tracings for Difference Engine No. 2 survive intact, but the designs show no sign of later revision. It seems that Babbage the perfectionist was at last satisfied, and the design was spared improvement.

But would building a Difference Engine vindicate Babbage at all? Would anything less than an Analytical Engine do? Bromley had anticipated this argument:

> Construction of the Difference Engine No. 2 would not only confirm the soundness of Babbage's logical and mechanical design principles in this case but would also lend conviction to the entire range of his designs for automatic computing machines.

So the successful completion of a working Difference Engine No. 2 would act as the touchstone of Babbage's standing. There was good reason for this assertion. Difference Engine No. 2 was designed between 1847 and 1849, well after Babbage had completed the major pioneering work on the more demanding Analytical Engine, and the elegant simplicity of Difference Engine No. 2 owes much to the mastery of technique he had already achieved. In terms of simple economy, Difference Engine No. 2 is three times more efficient than its predecessor – it requires one-third the number of parts for the same calculating capacity. There is another important link between the Analytical Engine and Difference Engine No. 2: they share the same apparatus for printing and recording results. A working Difference Engine would therefore provide strong evidence in favour of the logical and mechanical feasibility of the more ambitious Analytical Engine.

Bromley's proposal concluded with an understated challenge:

> The completed Difference Engine No. 2 would stand as a tribute to the forefather of the modern computer and one of the most ingenious mechanicians of the nineteenth century. It should be finished in time for the 200th anniversary of Babbage's birth on 26 December 1991.

Apart from arguments as to why building Difference Engine No. 2 would be a valid vindication and commemoration of Babbage's work, it was simply the only one of the Engine designs that stood the slightest chance of completion in time for the bicentenary of Babbage's birth. Bromley's proposal leapfrogged little details like securing approval, not to mention finance. But he had thrown down the gauntlet, and it was up to us to pick it up. There were six years to go and counting.

Bromley submitted the proposal to Margaret Weston who, before committing to the project, wanted a clearer idea of what constructing an Engine would entail. Bromley was due to return to Sydney to resume his teaching commitments, and he was indispensable to this exercise. Within days of his first appearance on my doorstep a meeting was hastily called in the Science Museum Library so that Museum staff could evaluate the technical feasibility of the Engine construction with Bromley as our guide.

On 29 May 1985 I strolled with Bromley past the Imperial College bell tower and across the campus green to the library, which housed the archive. I was about to see the original drawings for the first time. Bromley was matter-of-fact about the historical significance of resuming Babbage's abandoned venture. He had spent hundreds of hours poring over the drawings, and they were no longer objects of reverence or awe. I scampered alongside him, chatting as we went. By the time we reached the conference room on the second floor of the library, the rest of the group were already gathered. The original Babbage drawings were laid out on long tables before us. Present were the archivist, David Bryden (who had been Bromley's library host during his study visits), and the head of the Science Museum's engineering workshops, Mike Ball, as well as a senior staff member with mechanical engineering background. The five of us peered at the large, heavy, slightly yellowed sheets as Bromley began to explain the workings of the machine. If there was a moment that marks the genesis of the twentieth-century sequel to Babbage's bitter saga, then this was it.

The drawings are intricate and detailed. They show a machine eleven feet long, seven feet high and eighteen inches deep, with eight columns, each with thirty-one figure wheels. As Bromley talked us through the main features, Babbage's intentions became dimly intelligible. At this point Mike Ball expressed the first misgivings about the soundness of the design. Babbage shows the great Engine being driven by a large crank handle which the operator would turn in much the same way as a motorist would crank the starting handle of a car before the advent of electric starter motors. There is no suggestion in the drawings that any form of engine, steam or otherwise, would be used as a source of motive power. Ball, confronted with the tiers of meshed figure wheels, doubted whether a human operator, however determined, would be strong enough to turn the handle. He was worried that cumulative friction might make the machine inoperable. Privately, I was impatient with the suggestion that Babbage could have made such a fundamental mistake. But Ball was there to offer a professional engineering view and his concerns would need to be heeded. He might, after all, be right.

Although the drawings comprise a comprehensive

description of the Engine, they are insufficiently detailed to be of direct use in a workshop. Babbage did not specify what materials parts are to be made from, the method of manufacture, or the finish required for a particular external appearance. The parts are drawn to scale (some at quarter size) and the nominal size of parts is therefore known, but no dimensions are given. Crucially, the drawings do not specify the precision with which parts are to be made. This information is essential, both for machining the parts and for the historical objective of proving whether or not the Engine would work if made to the precision achievable in Babbage's time. For a modern workshop to make the parts, new drawings would be needed detailing each feature of each component – dimensions, choice of materials, finish, method of manufacture and the precision with which each part was to be made.

The issue of manufacturing precision was crucial to the build. Had Babbage proceeded beyond design to manufacture, the practice of the day would have seen him collaborating with his engineer to resolve such issues informally. Workshop practice has changed, and modern engineering workshops need fully specified piece-part drawings as much for engineering purposes as for contractual accountability. Before the Science Museum team could contemplate cutting any metal new drawings would be needed.

The lack of manufacturing detail provided by Babbage was not an insuperable difficulty. We had Bromley's specific knowledge to draw on as well as his broad perspective on Babbage's design style. Some of Babbage's preferences are explicit. He mistrusted springs, for example, and confined their use wherever possible to retaining a moving part in a particular position or biasing it against a stop, always in compression. In other situations his methods are less clear, and Bromley would be able to help resolve such issues by making comparisons with Babbage's other Engine designs.

The team also had the expertise of Michael Wright, the Science Museum's curator of mechanical engineering. Wright's knowledge of nineteenth-century hand tools, machines and workshop practice is encyclopaedic. The history of tools and workshops is not only his profession and vocation but his private passion. He has his own collection of period treadle lathes, including one by Joseph Whitworth, the man who gave his name to the first standardised screw thread and who worked for Clement on Difference Engine No. 1. To enter the basement workshop in Wright's home is to step into the world of Clement and Babbage – tools, jigs, partially restored planing machines, lathes with leather drive belts, cutters, oil-stained wooden benches and storage drawers. Wright is steeped in the nineteenth century, and in seeking his advice about materials, finish, methods of manufacture and period practice we would get as close as we could get to knocking on Clement's workshop door in Lambeth to visit the master. Wright was not present at the first meeting, but as the project developed he would be drawn in more and more.

Bromley and Wright, who were friends and colleagues well before the Engine project began, had little patience for the endlessly repeated thesis that the precision with which parts could be made in Babbage's day was the critical factor in Babbage's failure to realise a physical engine. Their own knowledge of nineteenth-century workshop practice told them that precision was not an issue. They had performed measurements on the parts of Difference Engine No. 1, and had already established that Clement was able to make repeat parts that differed from each other by no more than two thousandths of an inch. By quantifying the precision to which Clement was able to work, Bromley and Wright confirmed what they already suspected – that achievable precision was not a limiting consideration. But this had yet to be put to the test.

Until the 1850s there was practically no standardisation in manufacture in the machine-tool world. Each workshop had its own lathes, many of them custom-made, and the master screw from which screw threads were cut usually differed from lathe to lathe and workshop to workshop. The automatic production of near-identical parts in quantity was decades away. Instead, parts were made one at a time by a combination of machining and hand fitting. The understanding between the designer or inventor and the machinist, whether specified explicitly on a drawing or not, was that parts would be 'made to fit'. A part would be trimmed or tweaked by hand at the discretion of the mechanic until it functioned correctly.

The issue of how precisely each part was to be made was crucial to the mission of the build. If we succeeded in building an Engine, it would serve to establish that Babbage could have done so only if no part was made more precisely than could have been achieved in the middle of the nineteenth century. The project needed to be able to defend itself credibly against the charge that the twentieth-century team had made the Engine but Babbage could not have. Nowadays it is standard practice to indicate precision in terms of a 'tolerance' – the interval within which the dimension of the part will fall. I was dogmatically insistent that this element of the authenticity of the machine should not be compromised. Unless it was possible to quantify manufacturing tolerances, I argued, how would we know that nineteenth-century limits of precision had or had not been exceeded?

I returned time and again to Wright in the months that followed to press him for a range of tolerances to which Babbage would have worked. 'Tolerancing is a red herring,' Wright would say, again and again. He was patient as he explained to me each time that to talk of tolerances in the mid-nineteenth century was an anachronism, and that parts could be made to practically any required precision in Babbage's time. But I felt it was essential to be able to claim that we had been scrupulous in making no part more precisely than Babbage was able. Badgering Wright was an essential part of my education, and Wright was a generous and forbearing tutor. I finally accepted that the absence of engineering standards in the first half of the nineteenth century was an intractable fact and that, provided we kept a weather eye on the figure of two thousandths of an inch as a nominal limit of precision, we would not be compromising the exercise. The big unknown remained: would the Engine work?

The meeting of May 1985 had the desired outcome. While we were far from convinced that the machine could be built, we had found nothing obvious that told us that it could not. The collective view confirmed Bromley's original recommendation – that as a first step a small section of the Engine should be constructed as a testpiece before committing to full manufacture.

Babbage had taken the same cautionary path. From time to time he made experimental assemblies to verify the workings of critical elements. In the early 1830s he also made cardboard cut-outs of complex gear shapes to model their action. Constructing a trial piece would allow the team to verify the basic logic of the machine and resolve issues of precision and turning force that might be needed to operate the full machine. I was also aware of the strategic value that such a piece would have in promoting the project and raising the funds required for a full construction. It would serve as a psychological stepping stone to convince the cautious doubters. Dame Margaret agreed, and approved the scheme. We were finally under way.

Chapter 13

THE TRIAL PIECE

It certainly would be unsafe to rely upon the design being really complete, until the working drawings had been got out.

C. W. Merrifield, 1878

The section of the Engine Bromley chose for the trial piece was the arithmetical element at the heart of the machine. In computational terms the task it would perform is trivial – a school-room sum. The desktop trial piece would add a two-digit number to a three-digit number and give a result correct to three digits. But more importantly, it would verify a crucial feature of the original design – the ability to carry tens. The mechanism for the carriage of tens is the most subtle and beautiful in the Engine and is repeated over and over in the complete machine. So though the arithmetical ambitions of the trial piece were modest, it was vital for us to see that it worked before we could embark on the grand project to construct the whole machine. Since we would not have the benefit of the elaborate control system for the whole Engine, we would need a way of driving the trial piece to execute the various motions required for the calculation. Babbage and Clement had done exactly this in 1832 when they assembled the demonstration piece for Difference Engine No. 1.

To build the trial piece we would need detailed engineering drawings that specified the size and shape of each part. The specification of the trial piece needed input from Bromley. He was now back in Australia, but arranged to return to London for two months at the end of 1985 to supervise the production of working drawings. With funds sanctioned by Margaret Weston I hired a draughtsman to do the mechanical drawings under Bromley's instruction. Throughout January 1986 we worked against the clock. A man in a waistcoat and a hired draughtsman bent over the drawing board, pressing ahead with an Engine design – a scene projected from the past. With end-of-year festivities, a grim winter and the general inclination to hibernate, it was a bad time of the year to be piling on pressure. But enough was accomplished for Mike Ball in the Museum's engineering workshop to have a crack at estimating what would be involved: thirty-one working weeks, subject to the usual disclaimers for the reliability of the estimate which were, as it turned out, well founded. The drawings were a major leap forward, and everyone felt we were truly on the road.

I did a check against the calendar, remembering that Bromley had hazarded a figure of about six years to build a full Engine in time for Babbage's 200th birthday on 26 December 1991. I wrote to Bromley, now back in Australia again, and asked him for the basis of his estimate. He replied: a year to complete the trial piece, a year to produce fully specified drawings, a year to manufacture parts, and a year to assemble and test – with a year of contingency time, five years in all. So the first step for us was the completion of the trial piece in the first year. Ball's estimate indicated that one Science Museum workshop technician could meet this target with time to spare. In February 1986 it all looked rosy.

Optimism was misplaced. It took nearly four years to complete the trial piece. There was no single event that made a nonsense of a perfectly reasonable engineering programme, and I imagined Babbage nodding knowingly.

National politics played its part. This was Margaret Thatcher's Britain, and 'entrepreneurial Conservatism' was the new religion. Since she had become Prime Minister in 1979, much of the country had embraced the values of an enterprise culture, and the new rhetoric damned the years of industrial decline and the inefficiencies of state-controlled industry. The Museum was not immune from the cultural changes in public life. In March 1986 Dame Margaret Weston retired after thirteen years as director. She had risen from within the ranks of the Museum, and her leaving severed continuity with the old world. Neil Cossons, a vigorous museum professional, took up the reins. Entrance charges were levied, and a Marketing Division was created. 'Visitors' became 'customers' who needed to be wooed to part with admission fees. Change was no longer a transition between states of equilibrium but a permanent condition of corporate life. Constant redecoration became the norm. Power was transferred from keepers and subject specialists to the new managers. In the turmoil of reorganisation, projects became battlegrounds as senior staff jostled for purchase on the slopes of the new political landscape.

Those on the Babbage project, myself included, were too far down the line to be involved directly in the power politics. But the thunder from above was unmistakable, and the tensions filtered through to the ranks below. A usually congenial boss walking straight past you after a hellish meeting with his peers. No-notice cancellation of departmental meetings because of some crisis at the top table. Normal curatorial life cowering in the corner while the Titans battled it out. Each change of management meant new allegiances to be forged, delays, resubmissions, lobbying and pleading. Continuity on the Babbage project was lost in the upheaval. The redistribution of controls and financial resources meant that I would have to rebid for the project simply to revive it.

The new lines of authority were still obscure, so I bypassed a few tiers of the hierarchy and secured a meeting directly with Cossons. He gamely came at the appointed time from his plush director's suite to our scruffy departmental office. He always welcomed the opportunity to walk about the building, especially to staff quarters. These trips were part of his campaign to purge landings and corridors of abandoned furniture, long-forgotten filing cabinets, and the miscellaneous debris that accumulated in the age before tidiness and efficiency became one.

On 5 November 1987 I made my pitch to Cossons. I argued the historical importance of the venture, its research and educational benefits, its value as a visitor attraction and the lure of publicity for the Museum. I tried to clothe my zeal to build the engine with definable benefits to the new order – publicity, scholarly prestige and, well, publicity. I was in mid-flow and barely a third of the way through the prepared presentation when Cossons stopped me dead. 'Enough,' he said, with a dismissive sweep of his arm. That's it, I thought, we're dead. 'I'm convinced and happy to take the risk,' he said. 'Get on with it.' Meeting over. This was not the last time he would gamble on the project.

We now had the second approval to start the trial piece. The work was to be done in the vast well-equipped workshops that stretch the length of the Museum's basement. In their heyday the woodworkshop, paintshop, engineering workshop and foundry boasted a staff of 120 skilled workmen. Numbers had fallen to about a quarter of this as exhibition construction was contracted out and fewer and fewer exhibits were custom-made. Workshop staff were habitually over-stretched, and competition for technicians' time was settled by the decibel level of the bidder. With Cossons' authorisation behind us, a modest departmental budget was allocated for materials and Ball assigned an in-house workshop technician who had shown interest in the project.

Trouble was not long coming. It transpired that Bromley and the draughtsman had run out of time during January 1986, and the draughtsman had carried on alone without his mentor. For reasons that remain buried in the labyrinth of departmental politics, the draughtsman's contract was terminated with no notice by someone higher up. I had my suspicions about the motives for this but was never able to confirm them. The upshot was that the drawings were left unfinished. While they were adequate for Ball's rough estimate of build time, they were in places insufficiently detailed to be used in manufacture.

In the rush to produce the drawings no record had been kept of the reasoning behind design decisions, and I found it impossible to recover the rationale behind many of the features. The draughtsman had gone and Bromley was 10,000 miles away. With dodgy drawings, progress was halting. Queries from the workshop were frequent – omitted dimensions, materials, precision, and methods of manufacture. The project began to flounder. It was clear that the work needed daily supervision and someone constantly on hand to resolve difficulties, monitor progress and liaise with Bromley in Australia. My own curatorial duties to the computing collection and a demanding and ambivalent departmental boss did not allow me the necessary level of daily involvement. Peter

Turvey, a member of the curatorial staff in the Engineering Department, and weekend marine engine hobbyist, was assigned to co-ordinate the production of the ailing trial piece under my direction. Turvey, duly briefed, applied himself to the workings of the mechanism and began a faxed consultation with Bromley.

Bromley had already identified a range of oversights and errors of varying seriousness in Babbage's original drawings. As we delved further we found others. There is at least one case of a completely redundant mechanism: a long, beautiful, helical arrangement of arms which performs no function at all. At the other extreme, an absent mechanism made the last column of numbers vulnerable to derangement for a short time during a calculation. We were surprised, since Babbage had gone to extraordinary lengths to ensure the absolute integrity of the Engine's results. We also came across strangely shaped parts whose function we only partly understood.

But there was one flaw which dwarfed all others. The heart of the Engine and its most elegant feature is the mechanism for carrying tens. This mechanism ensures that if a figure wheel exceeds '9', a '1' will be added automatically to the tens wheel above it. Each figure wheel has a mechanism which 'remembers' whether the wheel has overflowed. The design of the mechanism is such that it will work in one direction only, when the number increases and passes from '9' through to '0'. In what seems to be an astonishing blunder the arrangement drawn in Babbage's original design drives the wheel in the wrong direction so that if built it would foul and seize the engine. The principle of the carry mechanism remains valid, but the layout as depicted in the drawings is unworkable. The error is so fundamental that the possibility that it was deliberately left as a precaution against industrial espionage still remains a credible explanation.

The errors precipitated a curatorial crisis. If the Engine was built as drawn by Babbage we had no expectation that it would work. If the project ran to completion the outcome would be an intriguing piece of Victorian sculpture, but the venture would prove nothing about whether Babbage's Engines were viable or not. If, on the other hand, the design was modified, in what sense could the new Engine claim to vindicate Babbage's work?

The way out of this impasse came when we realised that it was a mistake to see Babbage's design drawings as some sacrosanct entity of unimpeachable perfection. The attempt to build the Engine was not the physical realisation of an abstract ideal embodied in the drawings. It was instead the resumption of a practical engineering project that had been arrested in 1849 and remained in limbo for 140 years. Had Babbage taken the Engine off the drawing board to the next stage of construction, these design issues would have become evident to him. There would have been no way of avoiding them.

The question I now had to answer was this: how, given the tools, practice and precedent of his day, would Babbage himself have solved these problems? If we wanted to produce a working Engine we were going to have to modify the designs, but we would need to do so with uncompromising regard for historical integrity. We would have to analyse each error, seek out its likely source and come up with solutions that were consistent with practice in Babbage's day. Only then could the project hope to shed light on what Babbage himself might have achieved.

Over two years had passed, and the ticking of the clock was becoming louder. According to the original schedule the trial piece was now already a year overdue and there was still practically nothing to show except some apparently unusable drawings, a few forlorn bits in the workshop and a rather miserable technician.

In late October 1987 we abandoned the first trial piece. Two and a half years into the project and with only three and a half years to the deadline, we were going to have to start all over again.

The immediate challenge was to find a solution to Babbage's design blunder – if blunder it was. Although Bromley had spotted the problem, he did not attempt a solution in the first trial piece which, if built as drawn in the rushed circumstance of January 1986, simply would not have worked. Turvey and I racked our brains. Had we fully understood Babbage's intentions? Was he using some ill-understood form of subtractive arithmetic? Had the draughtsman simply made an error in the layout? Or was it really an anti-theft device to confound would-be copyists from building the machine from stolen drawings? Babbage did not draw any of the designs himself. He would have described to his draughtsmen the concept of the mechanisms, made rough sketches, and briefed them on what he wanted. Without records of this informal collaboration it is impossible to know how closely supervised the drafting process was and to what level of scrutiny the final drawings were subjected. After a respectable time spent agonising, the conclusion we reached was the least glamorous of the options – that the blunder was a drafting error either unnoticed by Babbage or, if noticed, one that remained uncorrected.

Solving the problem could not be postponed. We presented Bromley with our speculations and tentative conclusions, and he confirmed that in his view the faulty layout was an error. He went further and presented three possible solutions, all of which required modifying the design. One solution involved altering the sequence in which the Engine performed a calculation. By adding an extra procedure to the sequence the problem could be overcome, but at the price of extra time taken for the machine to complete a calculation. Babbage had gone to extraordinary lengths to shorten the cycle time of all his Engines. Economy of execution time was an obsession, and it would be uncharacteristic for him to incur a time penalty if there was an alternative. A second solution was to use subtraction as the basic operation instead of addition. Bromley's knowledge of techniques Babbage used in the Analytical Engine and elsewhere led him to believe that Babbage may well have adopted this option had he been confronted with the difficulties we now faced. There was a third way. All the carry mechanisms drawn by Babbage are right-handed, which is to say the layout of the parts is not symmetrical around a vertical line but leans to the right. A solution proposed by Bromley – and apparently hinted at by Babbage himself – was to make alternate mechanisms handed the opposite way. The solution effectively involved making alternate mechanisms in mirror image. Which solution was truest to the intentions of the nineteenth-century originator? Who could say for sure? We opted for the mirror solution as the one that would entail the least deviation from the original design. We now needed new drawings for a new trial piece to explore and verify the implications of left and right handing.

I began casting around for a seasoned mechanical design draughtsman for hire, with no success. The project was at a standstill. Then a hand from the past tapped me on the shoulder. Some thirty years earlier IBM had become enamoured of the Babbage Engines and commissioned replicas of the partially completed assemblies that survived from Babbage's time. The company that undertook this work was Rhoden Partners Ltd, in Acton, West London. There were some leftovers from the IBM replica project – a motley collection of casts and patterns, gear blanks, odd levers and paperwork. Out of the blue,

Gunther Wittenberg, a long-standing director of Rhoden, contacted me to see if the Museum was interested in retrieving the remains of the long since finished project. Here was a company that had direct experience of dismantling Babbage Engines, producing manufacturing drawings, making parts and assembling calculating machines. Its strengths were in engineering design and mechanical drafting – exactly what the Engine project needed. I visited Rhoden and took delivery of the material. As I left the building at Acton, I wondered what Babbage would have made of a company which, when it wasn't making replicas of parts of his calculating engines, designed wheel clamps and prototype machines for folding chefs' paper hats.

To make any progress, Rhoden's engineers would first need to understand the essential mechanical function of the Engine and adapt the design to mirror-image the carry mechanism. Late in January 1988 Turvey and I drove west from South Kensington the few miles to Rhoden's works in Acton. We left the car in the makeshift car park and went up in the goods lift. We traipsed through a large, stark, open-plan drawing office while being covertly inspected by draughtsmen and engineers from behind their drawing boards, and were ushered into the unadorned manager's office for the first briefing meeting.

Already seated at the cramped table was Reg Crick, a Rhoden design engineer, who would be assigned to the project. Crick, then close to sixty, was gentle, patient, and had an openness rare in someone who had survived the insecurities of a life in engineering design – the sudden scrapping of projects after years of work, the fluctuating financial fortunes of companies in engineering development, and the general wear and tear of professional life. His quiet modesty and his methods of working had an even pulse that never wavered.

The meeting went well, and Rhoden was duly contracted to

produce fully specified manufacturing drawings for the second trial piece. Crick set about the task of mastering Babbage's design and its known difficulties, especially the challenge of the layout error in Babbage's original drawing. He found that the solution using opposite-handed parts had unforeseen implications. Because the gear wheels in Babbage's design are not drawn symmetrically around a centre-line, a mirror image of the mechanism would produce a tooth where formerly there had been a gap. To remedy this he would have to rotate parts of the assembly through 2¼ degrees and carry this adjustment through the whole specification. Testing whether the mirror-image layout would work became the main purpose of the new trial piece. Issues of clearances, tolerancing and torque, which had so concerned us before, took a back seat.

As he waded in deeper, Crick started asking questions to which we had no clear answers. What materials did we wish to use for this part and that, brass, bronze, steel or iron? What style of font did we want for the numbers '0' to '9' engraved on the figure wheels? Did we want the parts to be cast or machined from solid material? Babbage had left us in the dark, and finding solutions would take time. Crick's questions were uncomfortable reminders of what lay ahead if we progressed to a full machine. But while the answers would be crucial to the historical integrity of the whole machine, there was no point in spending time on them until the trial piece confirmed whether the modified design would work. It was clear to me that materials used for parts were largely irrelevant to testing the validity of the mirrored mechanism – if it didn't work in steel, why should it work in brass?

But there was one question which plagued us more than most, and it demanded an answer: what shape should we make the gear teeth? The question was not a pedantic one and we were convinced at the time that the answer was essential to whether or not the Engine would have functioned in Babbage's day.

We have seen that the full Engine has hundreds of gear wheels, called figure wheels, organised in eight columns about three feet high. The wheels turn on a long shaft which runs vertically through the middle of the stack. The Engine adds by meshing the figure wheels in adjacent columns and turning them by an appropriate amount. During a calculation nearly five hundred wheels are meshed and driven, and it is here that the shape of the gear tooth comes into play. When one gear wheel drives another the wheels tend to be pushed away from each other, and the force of the thrust depends partly on the geometry of the gear tooth. The more efficient the design of the gear teeth, the lower the force trying to press the gears apart. So, as the gears turn, the shaft inside the column of gears will flex to some extent, however strong and rigid it is.

Since no one before Babbage had ever proposed using a shaft long enough to hold thirty-one number wheels, it seemed crucial to know whether Babbage could have made gear teeth that would not strain the shaft beyond the limits of correct operation. There are several standard gear-tooth geometries that we could use – involutes or cycloids, for example. But to casually use a modern tooth geometry not available or known to Babbage would again lay the project open to the charge that we had proved nothing about what was realisable in Babbage's time.

A lot was known about tooling for cutting gears – hob cutters, fly cutters and the like. But no one seemed to know what tooth profile Babbage would have used. Babbage's drawings were of little use. Nowhere did they say anything about gear shapes, and magnifying the draughtsman's line-work on the countless teeth shown in the drawings would not help either, since the teeth would have been drawn to a shape convenient for quick and easy drafting rather than one that accurately represented an intended geometry.

Anthony Hyman had done some of the legwork over a decade earlier while researching his biography of Babbage. He had borrowed some of Babbage's original gears from the Science Museum and had made enlarged images of the surfaces, which the Museum analysed in 1975. But the analysis was inconclusive. It revealed that the gear shape was neither one thing nor the other but an odd hybrid of two standard geometries. There was no time to pursue this further, and for the meantime we gave up and asked Crick to specify standard involute gears.

As Crick delved into the designs he discovered new features, made suggestions and offered solutions to problems he found. The drawings for the second trial piece took two months to complete, and the large roll of semi-transparent master 'skins' was delivered from Rhoden to the Museum in March 1988. We immediately made blueprint copies for workshop use and stowed the originals for safety. The drawings were to a high professional standard. Dimensions, materials, tolerances, screw threads, typeface for engraving, and so on were all fully specified. Finally, after a costly false start, we could begin again from a secure base.

Three years had passed. There were only three years to go till the bicentenary. Anxiety was a constant companion.

Metal was now once again being cut in the Science Museum's basement workshops. But progress was maddeningly slow. There was no one particular setback, no one set of circumstances that could be identified as responsible for the slippage. The roof did not fall in. No one died. Busy as the technician seemed to be, time trickled relentlessly by with frighteningly little visible outcome. Unlike engineering in a commercial environment, deadlines in the Museum were 'soft'. Survival did not depend on productivity. The workshops enjoyed craftsmen with incomparable skills who invariably produced exquisite work, and the practice was to take as long as it took. Quality and perfection were watchwords, and deadlines were part of the language of degraded standards. It was a fine tradition. But there was the uncompromising date with the calendar – Babbage's 200th birthday looming less than three years off.

The new commercial climate played its part in the delays. With the introduction of entrance charges in 1988, the Science Museum felt moved to provide novel visitor attractions to justify taking money at the gate and to attract more visitors. A huge mill engine was to be steamed up on a regular basis for the delight of paying customers. Unknown to me, the technician working on the trial piece was seconded for training and public gallery duties, and would only be able to work on the trial piece part time. There was no penalty to an overrun on the Babbage project, but lost revenue from a non-working mill engine was a measurable disbenefit. All I could do was cajole, complain, harass and plead. My written exchanges with the long-suffering Mike Ball were feats of stoic humour and elaborate protest, but the frustration between the lines is unmistakable.

The Science Museum had so far supported the project by funding the production of the drawings by Rhoden, assigning workshop staff to produce the trial piece and paying for materials. Looking ahead beyond the trial piece, it was clear that undertaking a full Engine would be beyond the capacity of the overstretched facilities of the Museum's workshop. I also knew that the cost of contracting out the manufacture and assembly of a full machine was beyond anything the Museum could reasonably be expected to fund from the ever-diminishing annual Grant-in-Aid from the Government. The Museum was moving from a Government-funded public sector ethos primarily dependent on Grant-in-Aid into a mixed economy culture supplemented by earned income and sponsorship. The transition was not yet complete. There were as yet no professional fundraisers on the staff, and I knew that I would have to raise the money myself. But without the trial piece as a platform for fundraising I was fatally handicapped in the campaign.

Raising money is an uncertain business and time was running out. I needed the completed trial piece, so I resorted to threats. I asked Rhoden to quote for the manufacture of the trial piece. About £10,000 and eight weeks to delivery, came the reply. I wrote to Ball saying that if the Museum workshops would not commit to complete the trial piece in-house in a comparably short time I proposed to put the work out 'without further reference to the Museum'. It was total bluster. I had little idea where the £10,000 would come from. I knew that Ball was too decent to be ruffled by rhetoric and that he would do his damnedest to deliver what he felt honour-bound to do. Ball took up the challenge, and in August 1988 put seven additional technicians on the job. The deadline came and went. No trial piece. There followed the ritual exchanges of recriminations, breast-beating and new promises. But still no trial piece.

I needed some high-level leverage. A Trustees' meeting was scheduled for 13 October and I volunteered the trial piece for display in the hope of using the event to pressure completion. The format of these gatherings is informal. Trustees gather for coffee in the plush ante-room of the large book-lined Fellows Room before the board session is convened. Curators, usually wearing white cotton gloves for the proper handling of objects, stand behind tables with new acquisitions laid out and chat to Trustees and senior management who show interest. Manuscripts, freeze-dried genetically engineered mice, an electric chair, a Minitel terminal, a personal organiser: these are some of the acquisitions that have made the cut for presentation. This half-hour quarterly event is one of the very few opportunities for direct contact between curators and the governors of the institution.

The trial piece, still not fully complete, went on show at the Trustees' coffee session on 13 October. Turvey worked the pitch. His gloved hands moved over the knobs and levers, producing the lifting and turning motions, locking and resetting, raising and lowering – the dozen or so separate operations needed to perform a single addition. I watched the reaction carefully. There was polite interest and modest encouragement. I volunteered the piece again for the 1988 end-of year Trustees' Jubilee party, a big social event for the great and the good, at which a range of exhibits and demonstrations were to be offered as party pieces. Ball again juggled his schedules, mustered more staff by taking them off other jobs, and as usual soaked up the flak from those he had deprived.

By the first week in December the trial piece was operable. A few components were still missing, but it took its place at the sparkling event on 12 December. Turvey again manned the stand, his startlingly white cotton gloves – to protect the metal from skin moisture – dancing over the knobs and levers. Like a latter-day Babbage, he entertained the glitterati of London to the delights and wonders of mechanised calculation. Babbage had amused his guests in his drawing room and later sought public recognition and support for his own efforts by demonstrations of his trial piece for Difference Engine No. 1. Here, generations later, was a group at the Science Museum doing the same.

There was no more political leverage to be had from classy

social bashes. When the wine stains were washed from the carpet, the trial piece was still unfinished. I needed a new strategy to drive the project. I chose the press.

It was crucial to raise the public profile of the project, especially in the computing industry, before launching the fundraising campaign. I went to a high-street newsagent, counted the number of computer-related publications on the shelf, and, based on the shelf displays, estimated that there were about 140 related publications. In visual terms electronic computers are pretty nondescript, and the media were ravenous for something that did not look like another boring box. Here was a story with images of sumptuous antiquity, leather-bound notebooks, 140-year-old design drawings, eccentricity, genius and failure – qualities celebrated in English culture. And redemption – the great crusade to vindicate the misunderstood progenitor of the modern computer age. It made a terrific story, and I flooded newsdesks with articles, photographs and carefully doctored sepia prints of Babbage's design drawings.

Computers were hot news. As curator of computing I was repeatedly being asked by the press for my comments on the latest products and developments. I started dropping hints about the project. *Computer Weekly*, a large-circulation trade paper for the computer industry, had run a piece as early as 1987. MUSEUM PLANS TO BUILD A BABBAGE was the announcement. As word spread, coverage grew. *The Times* ran several items. One, taking up a full half-page in March 1988, reported on efforts at Imperial College to use modern computers to simulate Babbage's mechanism as a design aid. SUMMING UP SHEER GENIUS ran the headline. The gist of the piece was that modern computers were liberating dear old Babbage from his nineteenth-century mechanical chains. In December 1988 *Scientific American* carried the story of the Science Museum's efforts to vindicate the long-dead genius of computing by building an engine.

In the new market-driven world I saw the media coverage as payback for the support the Museum had given, as well as a down-payment on support we might still need. I was convinced that targeting industry was as important as demonstrating to the powers that be at the Museum, especially the new assistant directors, that the project would have a high publicity pay-off, as well as offering historical and curatorial value. I made absolutely sure that press coverage was seen internally in the right places.

Television took up the story. Science programmes on the lookout for news picked up on the *Scientific American* article, and the Canadian Broadcasting Corporation was one organisation that asked to film the trial piece in mid-February 1989 for a programme called *The Nature of Things*. But the trial piece was still unfinished. I was weary of the memo war and the elaborately coded ways of expressing rage in a non-confrontational culture. I went out on a limb and confirmed the shoot. I then presented the booking to Ball as an institutional commitment. It worked. Overtime restrictions in the workshop were lifted and this time, to everyone's exhausted relief, it was done. The original schedule had allocated one year to make a trial piece. It had taken nearly four.

We prepared the trial piece for the shoot by polishing the large brass plates which framed the mechanism. We also brought to a high sheen the forest of knobs and levers by which the device is operated by hand. The reaction was ecstatic. So impressed were the film crew that after shooting the talking-head interviews they rustled up a revolving platform, covered it in black cloth, and ran the credits past the resplendent object turning slowly in the background as though preening itself in the glare of the film crew's lights.

The camera crew's reaction revealed something curious about the psychology of viewing: if you want to convince people that something is made accurately, it helps to make it shiny. Equating fine finish with precision is probably a perceptual legacy from the traditions of instrument-making. Telescopes, microscopes, sextants and precision laboratory equipment were often machined from brass and bronze and then polished, so the bright finish has become associated in people's minds with accurate work. The polished trial piece looked like a precious ornament. The film crew and passing staff clustered around it in wonder. I watched the reaction, and mentally booked the piece into the photo studio the next day for full-frontal colour shots.

As well as having telegenic appeal the trial piece actually worked, confirming not only that Babbage's basic design of the calculating element was sound, but that the mirror-imaging technique was a viable solution to the layout oversight in the original drawings.

With the trial piece complete I put out news to the press. CUTS COST BRITAIN 100-YEAR LEAD IN COMPUTERS was the accusation by the *Observer*, in a reference to the Treasury cutting off support to Babbage in 1842 after twenty years of public funding with little result. ACADEMIC ROW OVER BABBAGE MACHINE, blasted *Computer Talk*, though what the row was about is difficult to establish. THE BRASS COMPUTER was the soberly informed half-page piece in *The Times Educational Supplement*. The *Telegraph* carried a photograph. The story was out. The race was on to build the engine by December 1991.

The photo sessions in the Science Museum studio yielded colour prints of startling appeal. The contrast between the grained steel and solid brass spoke of solid dignity, precision, historicism and deadly seriousness. The polished surfaces played havoc with reflections, and the in-house photographers spent days in the photo studio experimenting with lighting and analysing trial shots. In the months and years to come the trial piece was photographed on countless occasions for news items, at company anniversaries and for catalogues, but the Science Museum studio shots were never bettered. Perhaps there was something to be said after all for the system of salaried staff photographers for whom quality rather than time or profit is the priority. The studio work was a far cry from the world of freelancers, flash-guns, egos and invoices. The images were more than I had hoped for, and would do splendidly well for fundraising. The next thing we needed was what John Herschel had politely referred to as 'pecuniary means'.

Chapter 14

THE MONEY

Go and build your Engine.

John Gardner, ICL, 1989

The financial arrangements between Babbage and the Treasury were a mess – an open-ended gentlemen's agreement, the terms of which were never clear. The murkiness of their unrecorded commitments led to bitterness and recriminations, and the unpleasant business took fully twenty years to resolve with the ultimate outcome unfavourable to Babbage. If there was a lesson to be learned from Babbage it was to avoid any form of open-ended deal. We would have to build the Engine for a fixed price or not at all.

Back in 1986 Allan Bromley had hazarded a figure of £250,000 to build the whole Engine, including the printing section. He took £5,000 as a baseline – the amount paid by IBM in 1970 for making a non-working replica of the small experimental section of the Analytical Engine. He multiplied this by a factor of five for the increase in size and complexity and a further factor of five to reflect cost increases during the fifteen years that had elapsed, and finally multiplied by two to account for the difference in cost between a working and a non-working machine. Answer: £250,000.

As it turned out this figure was uncannily accurate, though for the calculating section alone, without the printer. Instinct, inspired guesswork and a fudge factor are well-tested engineering design aids. But the figure would be difficult to defend to sponsors needing to be convinced that the team knew what it was doing. A factor of two in the wrong direction and the project would be caught badly short.

To set a reliable sponsorship target we needed a fixed-price quotation from an engineering company, and to secure the quotation we needed sponsorship to produce the engineering drawings specifying each of the thousands of parts. The logic was entirely circular. There had to be some way to break the vicious circle of quotation and money. I appealed yet again for internal funding. The Museum had already been generous to the project: it had funded the draughtsman in late 1985, contributed to Bromley's expenses for various visits, and stoically met Rhoden's bill for the production of the second set of trial piece drawings. This time I bid for the money to pay Rhoden to estimate production costs. The project had moved from the Engineering Department to the new Public Services Division. I was now on a loose rein to one of the new assistant directors and the unclear accountabilities of a structure in transition afforded greater freedoms to pursue the interests of the project than had been the case before. The budgetary structures were still uncertain. All I needed was an AD's signature, and I was more than content not to ask who was footing the bill. Again the response was positive, and in November 1988 I commissioned Rhoden to estimate the cost and timescale for producing fully detailed drawings, tooling and materials costs, parts manufacture and the calendar time to complete a full machine.

Preparing the quotation required new levels of technical understanding. The trial piece was operated manually by lifting, turning, lowering and resetting the cluster of knobs and levers on top of the device. In contrast to this, the full machine would be completely automatic. All the operator would need to do was turn a large crank handle, and the control mechanisms – none of which were included in the trial piece – would automatically orchestrate the necessary motions for the engine to calculate the next value. The elaborate control system of cams, rockers, sliders, racks and pinions, bell-cranks, bevel gears, links and levers that animate the intricate movements had never been examined in close detail with a view to manufacture. The logic and principles of operation had been cracked by Bromley early on, but for Rhoden's estimates to be realistic we would now need to go into the fine detail and take account of production issues way beyond those raised by the small trial piece completed in the Science Museum's workshop.

The lesson from Babbage's unhappy fate was that unless you could produce the hundreds of near-identical parts in a credibly short time and at low cost, the world at large – and bank managers in particular – would lose patience. For a while we enjoyed wistful notions of recreating a period workshop as a visitor attraction, and making the parts in public view using contemporary nineteenth-century techniques. The lack of time before the bicentenary and the lessons learned from Babbage quickly put paid to such rosy notions. We bit the bullet and opted for modern manufacturing techniques, taking care that parts would be made no more precisely than was known from measurement could have been achieved by Babbage, though possibly by other means. We would weld where Babbage would have forged, but the weld would be dressed to resemble a forged join. This was not deception so much as a conscientious exercise in visual authenticity. Since a greater part of the message to visitors would be in the visual spectacle of the machine, the requirements of authenticity were better met if the engine looked the way it would have had it been built in the mid-nineteenth century. The operational integrity of the machine remained uncompromised – if, that is, it worked at all.

Choosing to use modern techniques raised new questions. Producing the hundreds of figure wheels by casting (pouring molten metal into moulds) would yield near-identical parts more cheaply, but with the dull matt finish characteristic of the technique. On the other hand, machining from rough castings or solid billets would give the wheels a bright finish. Since the broadside view of the Engine presented nearly 500 wheels in all, the finish on the wheels would be a major factor in the Engine's appearance. I needed to ask Babbage or Clement what they would have done. I did the next best thing: I asked Michael Wright. 'Bright machined finish' was his considered reply. There was a cost advantage to casting, and we had yet to raise the money for the build. It was not clear whether the additional cost of machining from solid would make the whole project unaffordable, so I asked Rhoden for a split estimate – one figure for casting and the other for full machining. We could then quantify the cost of period appearance. Perhaps there would be money enough.

Time was taking its toll, and we began to downscale our expectations. The printing section of the machine is integral to its concept and design, and the plan had always been to build the whole machine as a single project. The printing apparatus is vastly ambitious and at least as complex as the calculating section. Despite our fascination with the printer and a sickening reluctance to compromise the grand venture, we were forced to accept that there was no realistic chance of building both the printer and the calculating section in time for the bicentenary in 1991. Not only was there not enough time, but the cost would almost certainly put the sponsorship target out of reach. Babbage had done the same in 1832 when he neglected the printer in favour of assembling a portion of the calculating mechanism to prop up his flagging credibility. Rhoden was duly instructed to ignore the printer and estimate for the calculating section alone. If we succeeded with the Engine, perhaps we could attempt the printer in due course.

While Crick and the backroom boys at Rhoden were sweating over the estimates, I was puzzling out an approach to raise sponsorship. Fundraising is a terrible slog. It involves endless enthusing and trying to please people with all the dignity of begging. The previous year I had costed half-scale versions of the trial piece for sale as executive toys as well as full-size replica trial pieces to be marketed as collectors' items or extravagant boardroom keepsakes. But the marketing legwork for this idea was daunting, and by February 1989 there was no time for such piecemeal measures. It would have to be done in one hit.

I explored another scheme. The cost of each machine would be much lower if three were built at once. Why not make three machines, pre-sell two for the full one-off price and fund the third from the economies of scale? Perhaps an American or Japanese corporation could be tempted into a prestigious piece of engineering art for a grand lobby, or a benefactor found to purchase one for a museum? As part of the commission to Rhoden I asked for estimates for up to three engines.

Rhoden had the figures ready at the end of February 1989.

These were prefaced by the ritual caveats about cost increases for the unforeseeable. But the essential data were there. One engine would cost £201,000. Second and third machines would cost £153,000 each. There would be a cost saving of £23,000 if the five hundred wheels were cast instead of machined – the price of period finish. The crucial figure now was the build time. This was estimated at fifteen working months from the date of order, provided they could overlap drafting and manufacture.

The time frame seemed encouraging, at least at face value, but less so when examined more closely. There were two Christmas and New Year factory shutdowns in the lifetime of the project, and because Rhoden's own manufacturing capacity was limited they would need to subcontract much of the manufacturing. The time needed to secure quotations from subcontractors had still to be built in, plus some contingency for unforeseen setbacks. When all this was taken into account Rhoden's estimate for the build time extended to twenty calendar months – assuming all went well, which is something no one realistically expected would be the case. We would need to place an order by November 1989 – in eight months' time – to have any prospect of having a fully assembled engine delivered by mid-1991. This would leave five months to get the machine to work by 26 December, Babbage's 200th birthday. Despite the delays we were still in with a chance, though there remained inherent unknowns as to what we would find when we came to put the beast together.

I had eight months in which to raise the money, and I set about compiling material for a sponsorship appeal. The pitch emphasised the unique promotional opportunities for sponsors as well as the historic importance of the venture. It featured a collage of press cuttings, a print of one of Babbage's original design drawings doctored with a tea bag in the printing process to give it a period sepia look, a published article on the project, and a techno-porn colour print of the trial piece on the cover. The pack appealed as a matter of urgency for £235,000 – a sum cooked up to cover one engine and a contingency allowance. Now, who on earth to take it to?

Times were favourable. In the late 1980s economics analysts and the financial pages of newspapers were in a frenzy of conviction that the future prosperity of the nation depended on the coming 'information age'. In his best-selling book *Megatrends*, published in 1982, John Naisbitt had forewarned of radical societal change: 'computer technology is to the information age what mechanisation was to the industrial revolution'. As the 1990s drew near, the burgeoning sunrise industries promised to spin silicon into gold. The computer industry and software houses were booming, and the personal computer revolution was ramping up. Computer manufacturers were sitting on a gusher. Society at large was less certain, bewildered by conflicting reports. Sceptics maintained that threats of apocalyptic change were the result of new-age techno-hysteria, fuelled by a sensationalist press and the self-interest of the information technology industries, and that all the fuss was hype.

The issues were hot and debate was rife. In July 1987, in the heat of public interest and when the computer industry was still buoyant the Information Age Project (IAP) was launched. The ambitions of the project were to create an Information Age Centre in Reading, Berkshire, to the west of London. The Centre would combine a visitor attraction on computing in the 'Silicon Valley of the South East' with acres of trade exhibition space and a conference centre. The prime mover behind the project was Berkshire County Council, and its officers energetically enlisted the support of computer and telecommunications companies – British Telecom (BT) , ICL, Digital Equipment Corporation (DEC), Rank Xerox, Hewlett

Packard, Siemens Nixdorf – and a property developer, Speyhawk. The Council secured the Science Museum's participation from the start to lend educational prestige, to provide exhibition content, and as a referee and guardian of promotional neutrality. Neil Cossons, the Science Museum's Director, was on the IAP's Board of Trustees.

The first phase of the project was to be a new £4 million blockbuster computing and telecommunications gallery at the Science Museum. As the Curator of Computing I headed up the Science Museum team doing the research, formulating an exhibition treatment and devising novel interactive computer exhibits. I was also a member of the Information Age Project Committee, and spent hectic hours belting up and down between South Kensington and Reading for meetings. So in the spring of 1989, when it came to identifying whom to approach to fund the Babbage Engine, I already had some access to most of the big-name players whose chairpersons and managing directors served as trustees of the IAP. There was one exception. IBM, 'Big Blue', the dominant presence in the computer industry, at the height of its corporate fortunes, had refused Berkshire's overtures and would not play.

While I was putting together the sponsorship packs, I received an invitation to an IBM conference on 'Science and the Unexpected', to be held at the Excelsior Hotel near Heathrow. I was already working gruesome hours on two projects, quite apart from the Babbage engine, and the pattern of seven-day weeks that was to prevail for the next three years was already established. There was no time for conferences. In any case, they were frowned upon in the Engineering Department as not being real work, and regarded with suspicion as an excuse for a jolly. But here was an opportunity to lobby 'Big Blue', the most prosperous computer company in history. Guiltily I went.

The event was more than a conference. It was an extravagant convention of people of power and influence gathered as IBM's guests to hear speakers from the ranks of the most prominent scientists of the day. It was a high-powered, brilliantly staged and expertly compèred techno-fest designed to impress and astound. The conference dinner was on 9 March 1989, and Sidney Brenner, the renowned molecular biologist, was the after-dinner speaker. I found my table, took my place, and settled down to take stock of who I was seated with. I checked the guest list. My table hosted a variety of distinguished folk, and I began to wonder what I was doing there. I introduced myself to my immediate neighbour, Geoff Kaye, a senior IBM man, and asked about the rationale for guest selection. 'Ah! Nobody's told you. This table is for people being lined up as speakers at future events of this kind.' IBM had done some advance planning, locked onto the Babbage bicentenary in 1991, and was lining me up for two years' time.

Here was an opening. With some difficulty I reached under my seat for the sponsorship appeal portfolio, and proceeded to bend Geoff Kaye's ear. Warming to my theme, I told him about the proposed new exhibition for the Science Museum, the Babbage bicentenary events, and the Engine project and its price tag. By the time I was done everyone else had finished eating and I turned my attention to my untasted food. It was surely uncool to have seemed so eager. But Kaye was willing to take it further. I exhaled. Coffee arrived and we settled back to enjoy Sidney Brenner, who was about to take the stage.

Kaye was as good as his word, and a few weeks later on 26 April he and two IBM colleagues met with me at the Science Museum to discuss possible sponsorship for various bicentennial events. It quickly became clear that the only basis on which IBM would consider support was as sole sponsor. It was all or nothing. If they were to support the Engine project they would do so without association with any other company, and their name and their name only would be publicly associated with the venture. It was a successful formula for IBM which was at the time sponsoring a sell-out exhibition on the works of Leonardo da Vinci on London's South Bank. The exhibition, called simply 'Leonardo', featured priceless original manuscripts, newly made full-scale models of some of da Vinci's flying machines, and other of his wondrous contraptions.

There was nothing coy about IBM's promotional visibility. At every opportunity the posters, presentation and publicity linked the two names – IBM and Leonardo, evidently on first-name terms. However distasteful such unsubtle connections might be to those with sensitivities about the appropriation of culture by commerce, it was undeniably a classy exhibition featuring incomparable material, and IBM as sponsor cashed in heavily on its association with culture, antiquity and genius. The 'Leonardo' model made me uneasy. I sensed a danger that IBM would use its promotional muscle to hijack the Babbage project, and that the company's name would be shouted while Babbage's was whispered. To get the Engine built I might have to live with that. But there was also a more serious strategic dilemma.

The prospect of a single sponsor had its attractions. It would spare me the exhausting and relentless process of doing the rounds seeking and negotiating sponsorship from multiple sources. The snag was that tying the project exclusively to one sponsor would run the risk of losing the whole project if the sponsor withdrew. It was a trade-off between convenience and risk. If IBM withdrew, there would be practically no time to recover the situation. Despite my misgivings, the prospect of one clean deal was almost irresistible in view of all the other pressures.

There was a new urgency. A major international conference on computing was being organised for early July 1991 as part of the Babbage bicentennial celebrations, with the Science Museum as joint hosts. This was planned not only as a historical conference but one in which the greats of the modern computer age would review and assess the state of the art and speculate about what the next 200 years might hold for us. The conference reception would be a spectacular stage on which to unveil the working Engine. With the prophets of the modern computer era as draw cards, media coverage was assured. The problem was that we had been working to 26 December as the deadline for an operational Engine, and the conference was scheduled for 1–3 July. The available time had shrunk at a stroke by almost six months. With this added urgency I went uneasily down the IBM route and started pressing for positive commitment.

The relationship with IBM was frustrating. In the labyrinthine corporate structure it was impossible to find out who the decision-makers were. Names of remote invisible people were uttered in hushed reverence. The main players were constantly jetting off to important meetings. It was always the next mysterious meeting that was the critical one, and the venues were always exotic. The Babbage project was in competition for sponsorship with the tennis at Wimbledon. There were weeks of black holes during which those involved were unavailable. I pressed on, sending letters and information to the people I had been dealing with, as the weeks slipped by. There were no replies except telephone messages about the availability of IBM staff for future meetings, which were usually postponed.

In July 1989 I travelled to IBM's development laboratories at Winchester. The purpose was to meet the IBM team that would be assigned to the project. I packed the trial piece in its fine oak carrying case, lugged it to the boot of my car, grabbed some drawings, and headed out of the Science Museum car park. At Winchester Geoff Kaye took me up to one of a row of identical offices and introduced me to two of his waiting colleagues. The standard furnishings and the 'clean desk' policy, which meant that there were no papers, clutter, or stationery anywhere in sight, created an impression of unnatural sterility. I learned that day that IBM wanted management control of the construction and that Kaye's two colleagues, who were retired or about to be retired, would supervise the project as a reward for distinguished long service. They would need to understand the Engine, assess its feasibility and schedule its production. I trundled through a demonstration of the trial piece and paraded Babbage's drawings, flogging myself into bright-eyed zeal. It became obvious to me that, however well meaning the IBM staff might be and however expert they were in their own professional fields, when it came to Babbage's Engines they were at the bottom of the learning curve. My heart sank at the prospect of repeating from scratch over the coming months the massive learning process the Science Museum team had already been through. We had done all the preparatory work using qualified and skilled engineers. We had the knowledge and understanding. Rhoden had the management and the engineering expertise. All we needed was the cash.

As a supplicant in the imperial presence I refrained from asking why it was that IBM thought it had the expertise to manage a mechanical engineering project that required knowledge of nineteenth-century workshop practice. Instead, I wearily played along. In a less charitable moment I mused how it was that the most successful computer company in history had succeeded in reducing a grand historic crusade to a retirement recreation for two of its staff. If I could only get them to commit the money, perhaps we could work something out about their technical involvement.

While flicking ash off his corporate sleeve, Geoff Kaye spoke breezily of internal bids being made for up to £500,000 – half for the Engine construction, and half for 'internal support'. I pricked up my ears. This term had not cropped up before. It transpired that, as with 'Leonardo', IBM wished to use the exhibition to showcase their top-of-the-range computer products and software expertise, by producing sophisticated computer animations of the workings of Babbage's engines. 'Internal support' referred to the resources that would be allocated to develop the computer simulations for display to astound the world with IBM's advanced technology. In my mind's eye a picture was forming: Babbage's name shrunken in size, the engine relegated to an alcove and in the foreground great arrays of computer terminals with interactive computer graphics by Big Blue. The lure of wrapping up a deal overrode my misgivings, and I held my peace. Back at the Science Museum I had sets of drawings laboriously duplicated and circulated them to the IBM team so that it could begin its education.

In August IBM hinted that the bid for support would be made in Paris at the end of October 1989. This meant another three months of uncertainty. If the decision went against the project, we would have lost months during which I could have sought alternative support. I pressed for a 'bail-out' option whereby IBM would commit to a sum of £30,000 as the cost of withdrawal, in case they did not exercise their sole sponsorship option – an agreement that had bound the Museum not to seek support elsewhere. If IBM committed to the bail-out penalty and ultimately withdrew from the project, the sum would allow us to buy a reprieve during which we could commission Rhoden to proceed to the next stage while I sought alternative sources of sponsorship.

By way of encouragement I launched another wave of press coverage. BOFFINS BUILD BABBAGE'S BOX, yelled *Computer Talk*. JUST EXCESS BABBAGE? quipped *Computer*. BABBAGED, said the *Guardian*. MACHINE MAY PROVE HISTORY BOOKS WRONG, warned the *Telegraph*. BRINGING BABBAGE'S COMPUTER TO LIFE, intoned *The Times*. All referred to the prospective completion of the engine in 1991. *Nature* devoted a full page to the new eleven-volume collected works of Babbage's published writings and included a centrefold shot of the trial piece.

In the months that followed I wrote four increasingly desperate letters to IBM seeking a commitment to the bail-out arrangement first discussed by telephone. No reply. In early October IBM announced their quarterly profits. They were badly hit, and a message on my telephone answer machine informed me that IBM had withdrawn from the project. The gamble of going along with one sponsor had failed. I had the sense not to call IBM's sponsorship programme manager until the next day. Even then I had to struggle not to lose it. I explained that announcing the cancellation of the project would cause a feeding frenzy in the computer press and that I did not see how IBM's name could be kept out of any credible account of why the project was being scrapped. Silence. I pressed on and argued that the Museum had acted in good faith in not seeking support elsewhere and that IBM should feel bound to honour the bail-out proposal and donate the £30,000 that would allow us to buy some time to salvage the situation.

IBM, though doubtless genuine in its wish to make amends, was caught in a trap of its own policies: it would donate funds only to projects it could publicly associate its name with, and only lend its name if the association was exclusive. It could therefore not donate £30,000 in partial support of the Babbage project. But the appeal to decency seemed to work, and a creative solution was found. We would seek a suitable existing £30,000 programme that IBM could sponsor in its entirety and the Science Museum could release this amount internally for the next phase of the Engine project. We foraged for projects to fit the bill. Happily, there was the production of a new information brochure, and that did the trick. We struck the deal, and within twenty-four hours I had mailed the order to Rhoden authorising the next phase of design and draughting.

We had won a reprieve for a few months. It was 19 October 1989. It had taken seven months to raise £30,000. There were just twenty months to go to the scheduled unveiling and still £200,000 to find. Unless the balance of the funding could be secured within a matter of weeks, the prospect of completing a working Engine by 26 December 1991 looked bleak.

The obvious alternative source of sponsorship was the rest of the computer industry. Most of the key players were trustees and potential sponsors of the Information Age Project at Reading, and all were well disposed towards the Science Museum. But to approach individual companies would constitute a breach of the collegiality of the IAP trustee group, and the Babbage project would anyway be competing for the same funds as the Reading project. The only strategy I could think of was to make a bid to the collective group and, as it happened, the next quarterly meeting was just one week away.

Ordinarily I did not attend trustee meetings. Papers had already been circulated, so there was no time to have the Engine project included in the agenda. As luck would have it three curators, myself included, were scheduled to give a presentation on plans for the Science Museum's new Information Age gallery to the IAP trustees. I had no choice but to hijack the meeting under 'any other business'. I reckoned anyway that surprise was a good tactic. It would avoid being headed off by those who did not want funds to be diverted from their favoured programmes into some madcap scheme to build a wacky machine. Apart from a hint to Cossons on the way down to Reading by car, I had told no one what I proposed to do.

The meeting made a start on its scheduled business. My colleagues and I revealed some of the wonders in store for visitors to the proposed gallery, and were well received. We sat down. As the meeting ground on I became anxious that the catch-net AOB item at the end would be crowded out and I would miss the only opportunity to make the bid. I waited to make my move. Time wore on. The trustees were becoming restless. Drinks and snacks were waiting. Finally, the chairman tried to close the meeting. He reached for his pocket diary and said, 'As there is no other business, let's set the date for the next meeting.' He looked around the table, inviting suggestions.

It was now or never. I took a breath, got to my feet and cast caution to the wind. In a reckless burst I told how our sole sponsor had suddenly withdrawn and that time was now critically short. I made an impassioned plea for the historical importance of what we were doing and for the unique opportunity that the bicentenary presented. The group already knew about the project from repeated coverage in the press. I declared that if I did not leave the room with a clear commitment of support from the trustees as a group I was giving notice that come Monday morning I would regard myself as at liberty to break ranks and approach each of the companies present on an individual basis – regardless of their pledges or commitments to the collective enterprise of the IAP. I stopped. I stayed standing. I didn't know what else to say.

A soft, calm voice broke the silence. It was John Gardner,

Managing Director of ICL, a leading computer company in the UK. Without making eye contact with any of the other trustees, without any sign of collusion or agreement with anyone else, he said quietly, and directly, 'If we do not leave this room with an agreement to fund the Engine, it will not be you knocking on my door on Monday morning, but me knocking on yours. If this group does not fund the build, ICL will.' His unilateral unequivocal support was the cue for the others. They went into a quick huddle. Five of them came back with a pledge for £200,000 for the Engine and another £150,000 for a special bicentennial exhibition on Babbage and his work. ICL pledged the largest share, and Hewlett Packard, Rank Xerox, Siemens Nixdorf and Unisys pledged equal shares to cover the balance. We could breathe again.

The meeting broke up. While we were milling around the food and drink I went up to Gardner to thank him. He waved thanks aside. 'Go and build your Engine,' he said. Gardner asked nothing in return. In another country, in another culture, I might have been tempted to give him a hug. But this was England, and anyway we were each holding a plate of mushroom vol-au-vents.

Chapter 15

THE DEAL

I was ignorant of that which no human being could foresee.

Charles Babbage, 1834

With pledges of funding in place and Reg Crick crafting the new drawings at Rhoden, I shifted attention to the terms of the contract. Rhoden was reluctant to commit to building a working Engine for a fixed price. No one had ever built a Babbage Engine, and there were inherent uncertainties about whether the Engine would work, however exact the design and specification. On the other hand – especially given Babbage's own disastrous experience with a muddled open-ended arrangement with the Treasury – I was determined to secure a fixed-price deal that would protect us from cost over-runs that are a notorious feature of public works.

John Reid, a cheerful Scot newly appointed as the Science Museum's head of finance, took up the cudgels. Reid and Eric Parsons, Rhoden's financial director, embarked on some virtuoso bargaining. Both knew the game well and relished the process. It was knockabout stuff and each invoked the age-old ploys of bluff, counter-bluff and the secret bottom line. A compromise was reached. Rhoden need not commit to build a working machine but must commit to making parts exactly to the specifications in the drawings. Rhoden would need to make good at its own expense any shortcomings in manufacture. The accuracy of the drawings was entirely the Science Museum's responsibility and I undertook to sign off each of the hundred drawings as the official record of what was required. If the parts were made as specified and the machine did not work, Rhoden would have no responsibility to remedy the situation, if indeed it was remediable at all. To provide some protection to Rhoden, I proposed a contingency fund of £20,000 to cover needs unforeseen during the design phase by both parties, and Rhoden would have to rigorously justify any call on this reserve. The figure of £20,000 was capped as an absolute condition.

Everyone had assumed that once the design drawings were ready, Rhoden would be contracted to manage the manufacture of parts and then supervise the build. But in the anxiety over funding and the general urgency of the project, no one had taken into account that public sector procurement regulations stipulate that contracts above a certain value have to be put out to competitive tender, and that we would have to get three separate quotations for the 4,000 parts. Technically the Museum had no entitlement to engage any single party, Rhoden included, to undertake the Engine construction. The tendering process would take three to four months, and the process could not begin until all the drawings were complete. We would lose not only the extra months but also the time we had hoped to save by overlapping the start of the manufacture with the completion of the drawings. The inference was inescapable: going to competitive tender at this stage would scupper the prospects of completion by the bicentennial deadline. In my more paranoid moments I thought I detected a concealed satisfaction in some less than sympathetic quarters at this new obstacle.

John Reid scurried off to scrutinise the terms of the procurement regulations, and he and I got together to try to unpick the problem. I suggested that Rhoden had unique expertise that would make the project undeliverable by anyone else. Reid correctly argued that once the design drawings were complete, it was a straight manufacturing job that any number of companies could undertake. The companies manufacturing parts would not need a highly specialised understanding of how the Engine works nor would they need to go through the arcane process of interpreting Babbage's intentions. All they had to do was to make the parts as specified in the drawings. Rhoden's 'unique expertise' could not therefore be grounds for exemption from the requirement to go to open tender.

Reid suggested another tack – that we use the special provision for time-critical projects. If the time to tender jeopardised the delivery of the project, then in exceptional circumstances dispensation for a single-party tender could be granted. It looked as though we were in with a chance, but given the size of this contract and the high visibility of the project, the only person empowered to authorise the exemption was the chief accounting officer of the Science Museum – the director, Neil Cossons. I booked a slot with Cossons for the next day, 20 December 1989, and prepared the papers for signature overnight.

Even if we succeeded in convincing Cossons to authorise a single-tender contract with Rhoden there was still the question of financing the build. Rhoden would require a down-payment in advance for the subcontracted work.

Although the sponsoring companies had pledged funds, there was still no money in the bank. I prepared a second set of papers for Cossons' signature. This was a commitment by the Science Museum to provide bridging finance to fund the project until the sponsors paid up. This would entail a degree of risk for Cossons. The computer industry was showing some volatility, and the sponsors, however well meaning, could yet find themselves stretched come payday. For Cossons to provide bridging finance was tantamount to underwriting the costs of the project and accepting that the Science Museum would foot the bill if for any reason the sponsors did not. Babbage would surely have appreciated the irony of the Science Museum, which as a public body was funded by the Treasury, bankrolling his Engine to completion 146 years late. Cossons needed little convincing. He took the gamble and signed both documents with a flourish of broad black ink.

Back at Rhoden, Crick was generating a set of new drawings specifying each of the 4,000 parts. He measured each dimension of each part on Babbage's original drawing, scaled up where necessary to full size, and inserted the result on a new drawing depicting the part in all its engineering detail. The technical issues that had earlier been deferred under pressure had now to be faced and resolved. A host of new queries, anomalies and uncertainties had arisen in the meantime as Crick took the project to the last level of detailed specification before construction.

Some of Babbage's original drawings are difficult to decode, though the quality of the drafting is uniformly high. Views at different levels in the machine are drawn on top of one another on the same sheet. Overlaying multiple views in this way produces drawings of such density that there are occasions when details are impenetrably obscured. It was impossible, for instance, to establish whether the large cast frames that support the whole engine were made as single castings or assembled from sections. This was just one of many puzzles.

There are inconsistencies in the dimensions in some of the original designs. The selfsame part is sometimes shown differently sized on different plans. There are also omissions and incompletenesses. Three rollers, for example, which automatically ink the print wheels are shown suspended in space with no visible means of support and no detail as to how they are to be driven. The motions would have to be derived from other moving parts and the drive train designed accordingly. Although we had given up on building the printer for the meanwhile, I was determined that we make provision for it so that its eventual construction would not be prejudiced by neglect at this stage. There are also anomalies in the spacing of parts, the most glaring of which appears on the main drawing showing the Engine as seen from the front. This drawing is the most revealing of the Engine's overall structure and is also the most evocative of its overall shape. Eight vertical columns are shown, each made up of thirty-one separate number wheels. The gap between the columns is the same except in one instance where the uniformity of the arrangement is unaccountably broken. There is no functional reason for this, and the machine will not work if built to the scheme shown.

There were other puzzles. Lubrication was one. While grease and oil reduce wear and friction they also act as dust traps. Cleaning an engine of the complexity of Babbage's would involve dismantling it in its entirety and reassembling it – a mammoth task. Clocks, watches and locks are usually run dry for this reason. Babbage gives no written instructions on whether or not his engine is to be lubricated. In only one instance does he make any provision for lubrication, where he specifies an oil bath at the base of a heavy vertical camshaft. But it is clear from the drawing that once the mechanism is assembled access is completely obstructed, and it is impossible to inject any oil or grease into it. Perhaps Babbage had intended this as a 'sealed for life' bearing, innocent of the now recognised fact that in modern parlance 'sealed for life' is taken as a self-referential term in which 'life' means the life of the bearing, which in turn determines the life of the whole machine.

Crick spotted the problem at the design stage and devised a solution. He hid an oil bath under a large bevel gear at the top of the shaft and ran a concealed channel down the shaft leading to the bath below. Oil trickles down unseen to replenish the supply, and the whole arrangement is invisible to ordinary inspection. For the rest, we elected to lubricate the machine by hand using grades of oil and grease that do not stain bronze, and to provide drip trays on the floor to catch the oil as it drained through.

There was one persistent question that could no longer be postponed: what materials should be used for which parts? Inspection of the relics of Babbage's partial assemblies as well as expert advice from Wright and Bromley indicated that the use of gunmetal, cast iron and steel would be consistent with the period. Gunmetal, a form of bronze, was in common use in Babbage's day. Admiralty grade gunmetal is an alloy of 88% copper, 10% tin and 2% zinc. The physical properties of alloys are significantly affected by the addition of small quantities of other elements such as lead and phosphorous. How knowledgeable were Babbage and his contemporaries about the effects of such additions on durability, friction, brittleness and the workability of the resulting alloy? Would he have subtly adjusted the recipe in some way to suit the special needs of his Engine? If the modern construction used a grade of metal not available to Babbage then the integrity of the construction would again be open to question.

The problem lay less with the irons and steels than with brasses and gunmetals. Visual inspection of the surviving machines from Babbage's time was inconclusive: the ageing of the metals and the patina produced by surface corrosion make it difficult to tell by eye whether a part was of brass or bronze. I concluded that the only way of resolving the issue for certain would be to establish by analysis the composition of contemporary metals actually used by Babbage himself.

As luck would have it, the computer collection had several miscellaneous loose parts among the Babbage Engine relics. What we needed now was a favour from a friendly metallurgist with sophisticated test equipment. Across the green and past the bell tower is Imperial College of Science, Technology and Medicine, a university of the highest repute, which happily for us has a Department of Materials. Choosing four loose parts from Babbage's leftovers, Turvey and I went to visit our neighbours to beg expert advice and some sophisticated materials analysis. We were sympathetically received, but the news was not all good. To perform an accurate analysis it would be necessary to remove a small sample of the host material by sawing or chipping. Defacing an object in this way goes against the basic tenets of curatorial practice, which regards original artefacts as an inviolable part of the physical record.

An alternative was a non-destructive technique in which the sample was bombarded with an electron beam, and the X-rays given off were examined. This would give results no more accurate than 1%, but it would be enough to identify the class of material and its approximate composition. In curatorial terms even this technique was not entirely cost free. The corrosion products that had accumulated on the surface for over a century would mask the host material and give misleading results. A minute area on each sample would have to be rubbed with diamond paste to abrade the surface and expose the underlying metal. We had little option, and went ahead. The X-ray microanalysis revealed just what we had hoped: the composition of Babbage's gunmetals showed some variation but they were all within an accepted standard mix of copper, tin and zinc with small quantities of lead. Finding a modern bronze to match would now be easy.

The unresolved design issues that Crick had raised could no longer be put off. In January 1990 I scheduled a series of blockbuster meetings of an advisory team. By a happy coincidence, Bromley joined the group. He was on his way to Greece to work on an early Byzantine gearing mechanism found near the island of Antikythera, and would be in London as a visiting research fellow at the Science Museum for a few weeks.

The posse of curators, historians and engineers consisted of Crick, Bromley, Wright, Turvey and me. The five of us sat under fluorescent lights in the stark meeting room at Rhoden's premises for four full days of marathon consultations relieved only by tea, coffee and the occasional gobbled sandwich. We waded through the drawings of the 4,000 parts for the vast machine. We specified the materials for each part, which of the number of possible solutions would be most true to contemporary practice, and which method of manufacture would least compromise the visual and functional authenticity of the end result. Wright's vast knowledge of nineteenth-century tools and workshop practice was invaluable and we turned to him again and again. The work was gruelling and painstaking, and tempers frayed at times under the pressure of having to resolve issues on the spot with no comeback. The mountain of detail was oppressive. We argued through every issue from successive standpoints – cost, historical appropriateness and design.

At the end of the last meeting, on 17 January 1990, we slumped in our chairs, thoroughly wrung out. We had gone as far as we could. We sat in silence, numbed by the unrelieved concentration on detail, each of us in a private reverie of relief. Bromley was the first to stir. For no obvious reason he stood up and taped one of Babbage's printer drawings to the wall. We had decided months ago to abandon the printing apparatus, and the printer design had been excluded from the marathon consultations. We were mildly mystified as to why he should turn to the printer at this late stage, but too tired to rouse ourselves.

It was years since Bromley had first decoded the drawings. He started to explain the design. As he went on, his own sense of rediscovery of the intricacies and beauty of its conception and workings showed in the exhilaration with which he traced through the progression of numbers from the final column of figure wheels to the print wheels in the stereotyping apparatus. It was a bravura performance which lifted everyone out of the pit of grinding detail in which we had sunk. It was a relief to think in an unfettered way after long sessions of constrained and disciplined reasoning. A free-for-all ensued and we showered Bromley with questions. The session reminded us of our early wonder at the ingenuity of the machine and the tantalising uncertainty as to whether any of it would actually work – an abiding puzzle that had been temporarily buried in the numbing detail of the last few days.

By the time Crick had finished the new specification, Babbage's twenty-four original sheets had been translated into 100 large new drawings which fully specified the parts in all their detail. The agreement was that Rhoden would manage the production of the parts by subcontracting manufacture to other companies specialising in various processes – milling, casting, gear cutting, spring winding, case hardening, and so on. As the drawings came off Crick's drawing board they were sent in batches for independent quotations from three separate engineering companies – Pique Precision Engineering, Douglas Curtis and Specialised Engineering Products (SEP). Each company costed the manufacture of all 4,000 parts, and the massive job of mixing and matching the most economic combinations of supply was begun. Barrie Holloway, a Rhoden engineer, was in charge of this nightmarish task. With Crick and Holloway I visited the three companies to view their plant, get a feel for their operation, assess the reliability of promises and review delivery, so that we could judge how best to distribute the order between the three contenders.

The final quotations were ready at the end of May. The results were disturbing though not entirely unexpected: the cost had gone up from £200,000 to £246,000. The reasons for the increase were part of a familiar litany in the industry: the scarcity of specialist tool-making skills, inflationary cost increases, the unbudgeted costs of conducting the tender and cost overruns in completing the set of design drawings. I notified the would-be sponsors as well as Neil Cossons and the Science Museum's finance section, which was bankrolling the project against sponsors' still unrealised pledges. I waited nervously for flak, remonstrations, muscular censure. Nothing happened. It was all more or less expected. Anyone familiar with Babbage's history of spiralling costs was not the slightest bit surprised.

Meanwhile, the contract between Rhoden and the Science Museum was being thrashed out. Reid, Parsons and I re-engaged in our ritual arm-wrestling and converged on a set of mutually agreed terms. Signature was imminent, and a down-payment of £67,000 would be payable to Rhoden with the order. The quotations from the subcontractors were valid only for a fixed period, until Friday 8 June 1990 – just a few weeks off. If we overran this date without placing firm orders we would have to suffer a second round of tendering for new quotes. Apart from the delay the process would also incur cost increases. After a long haul over five years it seemed that we were nearly in the clear – drawings done, fixed price agreed, contract ready for signature and funding in place. We prepared to sign. Finally, we could look forward to cutting metal and beginning to realise the object of the whole fraught exercise.

The press, which was closely following progress, went public on the imminent signing of the contract. THE NUMBER CRUNCHER: A 19TH CENTURY FAILURE, ran the headline in the *Financial Times*. MUSEUM REVIVES GEORGIAN GENIUS'S TECHNOLOGY, sang *The Times*. Then, with no warning, Rhoden went bust.

What no one at the Science Museum knew was that Rhoden had been in trouble for months. While I had been fundraising, Rhoden had been running out of work and had kept busy working on the Babbage design drawings without payment. From October 1989 they had been on a four-day week, and in May 1990 the lease on their premises ran out. The workshop machines were put in store and staff worked from home. Reassured at first by IBM's interest and by the standing of the Science Museum as a client, they were confident that the money would be raised come what may, and they kept going on trust. I was not to learn this till much later. As it turned out, the lead time saved by this reckless commitment to the project made a critical difference to shortening the delivery time of the drawings. Knowing just how precarious the whole business of securing sponsors had been, I was dumbfounded that Rhoden's had opted to struggle on in hope.

Eric Parsons telephoned John Reid at the Science Museum on the morning of Tuesday 5 June 1990 and called an immediate meeting. Reid and I assumed that the purpose was to take delivery of the contracts for signature so that Rhoden could subcontract manufacture before the quotations expired in three days' time. The three of us met in Reid's office. Parsons, a voluble and hugely self-assured man, came in and sat down. He was shaking. He had come directly from his chairman to tell us that Rhoden, a specialist engineering company that had been in business for over thirty years, was going into receivership in two days' time. Despite the Babbage order their work had run down to an insupportable extent and the company had no option but to fold. Rhoden's staff did not yet know. He was on his way to tell them. He and they would be out of a job in forty-eight hours.

Reid and I were reeling from the implications of being deprived of Rhoden's management role as well as losing the one firm authorised as the recipient of a single-tender contract. More than anything, I was appalled at the prospect of losing Reg Crick, the engineer who had taken such commanding control of the design and specification of the whole machine. The project had come too far over the past five years for us to give it up now. We had to rescue the situation somehow. There was no time to cultivate a new relationship with any other company: learning time was prohibitive and the working relationship was irreplaceable. On the hop, I proposed to Reid that the Science Museum hire Crick and Holloway directly. The Museum itself would have to let the contracts for manufacture. I would oversee the construction and manage the project to completion.

By his visit and early disclosure, Parsons had behaved with scrupulous propriety and had pre-empted payment of the large advance which would otherwise have vanished into the unrecoverable assets of the liquidated company. Parsons went off to Rhoden, an honourable messenger of doom. He had a message for Crick and Holloway that was to be kept from the rest of the staff: they were to start work at the Science Museum on

Friday 8 June, the day after they were fired and the deadline by which orders had to be placed with subcontractors. Reid and I had just two days to square what we were doing with Cossons, secure the necessary approvals, arrange office accommodation, and prepare contracts and job descriptions for the two shipwrecked engineers.

Crick and Holloway duly showed up on Friday morning. Their relationship with the Science Museum had changed overnight as the Museum transformed from client to employer. Reid and I ran through their professional backgrounds, experience and standing, and negotiated a fixed-term contract starting with immediate effect. It was all done by noon and Reid, the cheery Scot, was in a mood to celebrate. To seal the deal he invited everyone for lunch at a local Italian restaurant, Piccolo Venezia, a few blocks from the Science Museum. I tried to protest. The deadline to mail the orders to the three contractors was the end of the day. We had a little over five hours to prepare the detailed parts lists, three sets of 100 drawings and the official orders. Worse still, the contract that had been negotiated over the past months was relevant only to Rhoden, and not to the subcontractors that would actually manufacture the parts, so the terms and conditions of the existing contract were largely unusable. I had only five hours to concoct a new document, and the prospect of eating up some of this valuable time in a restaurant gave me the heebie-jeebies. But the jocular Reid was persuasive and, not wishing to be a party-pooper, I succumbed.

Lunch was a protracted affair and Reid an entertaining host. I sneaked regular glances at my watch, each time calculating how long there was left before the Post Office closed at 5.30 p.m. But Reid's hospitality prevailed, and we did not get back to the makeshift office until 3.00 p.m. Immediately we launched into a frenzy of activity. Crick and Holloway attacked the mountain of paperwork hastily transferred from the now defunct Rhoden Partners. They duplicated parts lists specifying how many of each specific part each of the three main contractors was to make, and folded 300 large drawings into three separate packs for mailing.

I started writing the contracts from scratch, trying to foresee as many difficulties as possible and to provide redress in the event of things going wrong. The document covered defective manufacture, unauthorised overproduction of parts, milestones for stage payments, delivery and a host of other stipulations. In Babbage's day the tools and jigs used in the manufacture of parts for Difference Engine No. 1 were by law the property of Joseph Clement, the engineer hired by Babbage, even though many were made specifically for the Engine at Babbage's expense. I stipulated, by contrast, that all patterns, special jigs and paperwork were the property of the Science Museum and were to be returned when the job was done.

It was 5.15. The Post Office on Exhibition Road alongside the Science Museum would close at 5.30 p.m. The consignments must be date-stamped no later than Friday 8 June to beat the expiry deadline. Holloway salvaged some used brown wrapping paper from the bottom of a cupboard, trussed up the three bundles with parcel tape, scribbled the addresses and handed them over to me. I sprinted down the alley between the Science Museum and the Geological Museum to the Post Office, clutching the bundles to my chest. By my watch it was 5.28. The large oak doors of the Post Office were shut and securely locked. I walked slowly back, trying to catch my breath. Crick and Holloway had gone. I dumped the three parcels in my car and drove home. Maybe if I posted them first thing Saturday it would be good enough. It was.

Chapter 16

THE BUILD

All this is very pretty but I do not see how it can be rendered productive.

William Hyde Wollaston, 1822

The Engine was assembled in full public view on a prime site on the ground floor of the Science Museum in South Kensington. Visitors on the way to the Museum's galleries passed the Engine as it slowly grew. Barriers were constructed to enclose a secure area, like a small stage set, in which the two white-coated engineers, Crick and Holloway, worked on the machine. The cordoned-off area with its stainless-steel handrails became known to the team as the 'cattle pen'. Blue banners suspended from the high ceiling proclaimed: 'Babbage Engine Project', and a flip chart on an easel announced each day what work was in progress: 'Today we are assembling the cam stack' was an early message. Visitors were free to talk to the engineers. There were no rehearsals for the engineering work, and vile oaths and cussing were banned. Grazed knuckles, uncooperative parts, botches, disputes and clangers had to be dealt with in a manner inoffensive to the visiting public. It was all quite unnatural.

Selecting a site for the build was difficult. Everyone seems fascinated by work, and assembling the machine in public view would make a splendid visitor attraction. On the other hand, allowing visitors to talk to the engineers would inevitably slow things down. The timescale was tight, and even if all went well the uncertainty over what lay ahead was a constant anxiety. An insistent public might prove to be a source of irritation and even harassment to engineers working against the clock.

There was another factor to consider in the choice of site. No one knew if the machine would take kindly to being moved after it was assembled, assuming that we got that far. The Engine would be cross-braced with special stays to give rigidity to it during transport. Even so, if it was built any distance from its final display position we feared that stressing the framing during transit might distort the frame and affect critical sections of the mechanism. Uncertainty about the machine's vulnerability overrode concern about the risk of disruption from an inquisitive public, and we selected a build site only thirty yards from the destined display position on the same unbroken floor level. All that would be needed to transport the engine was some heavy-duty lifting gear and poles or skates to roll the engine from the build site to its place of display.

During the months since Rhoden's crash, specifications had been farmed out to three main contractors. They had in turn subcontracted to other specialists to complement their own in-house skills: pattern-makers, foundries and companies that specialised in gear cutting, cam profiling, case hardening, planing, brazing, welding and graining. Modern manufacturing techniques were used, and computer-controlled machinery produced the hundreds of repeat parts, the manufacture of which had so handicapped Babbage. Babbage had used a single component supplier, the engineer Joseph Clement, who made some 12,000 parts in eleven years. In all we used forty-six separate subcontractors who manufactured the 4,000 parts in under six months.

The first parts were delivered in late September 1990. These were two large rails eleven feet long, much like railway lines, on which the whole three-tonne engine would rest. Any excitement at the first delivery quickly evaporated. Sighting along their length showed that they were slightly banana shaped instead of straight. They had bowed during casting. It was a disastrous start since the assembly could not proceed until the rails were laid down. The terms of the contract placed the responsibility squarely with the supplier, and we would have been well within our rights to insist on having the rails remade. But this would hold us up for weeks. Crick and Holloway pondered the problem and worked out that if the rails were prised apart at one end there was just enough metal to elongate the holes at the far end for the fixing holes to line up. The offending rails were sent back to the contractor for milling and pressed into service a week later on return. It was a shaky and nerve-racking way to begin.

The media were still keeping tabs on the project, and I used the official start of the build as occasion for more fanfare. VICTORIAN NUMBER CRUNCHER FINALLY GETS PAST THE DRAWING BOARD, announced the *Sunday Correspondent*. 'These men are building the first computer' was the *Telegraph*'s caption to a photograph of Crick and Holloway taken fetchingly from inside the framing. 'Cast iron computer' captioned the *Financial Times* photograph of Crick peering intently at a bevel gear. LONG LIVE THE BABBAGE DIFFERENCE, exulted the *Guardian* with a photograph of Crick, Holloway and me frozen like waxwork dummies. Media interest extended to television and radio. Channel 4 used the cogwheel engine in a programme, *Little by Little*, devoted to nanotechnology – the topical new world of mechanical engineering on a molecular scale. Cable Network News and BSB Satellite news covered the start of the build and the historical ambitions of the venture. BBC1 Breakfast Time and NHK (Japan) followed suit. The BBC World Service and Radio 4 carried news of the grand happening and the prospect of assembling the first few bits and pieces.

Publicity was still strategically crucial. The sponsors' funds were tied up with the IAP project, and they had been unable so far to release money for the Engine. Publicity would serve as encouragement. Until the money was in the bank I resolved not to refuse any photo calls or media interviews, although they took time and interrupted the continuity of the work.

I was working inhuman hours and had to crank up another notch. I was still running the IAP computer gallery project and co-authoring a book for a five-part BBC series on the history of computing called *The Dream Machine*. At the same time I had set myself the task of reading everything Babbage published by working my way through the newly issued eleven volumes of his collected works. The reading progressed at nights and on weekends at the painful rate of five pages an hour. Babbage has a habit, both engaging and frustrating, of returning time and time again to issues that preoccupy him – and burying gems in unlikely places. The page-by-page scrutiny, cross-referencing and note taking, went on for eight months. I had also recently founded the Computer Conservation Society, a joint enterprise with the British Computer Society, to restore to working order some of the historic machines in the Science Museum's computing collection. The Society was set up from scratch and staff needed to be recruited, office and workshop accommodation found, and working practices established for the cluster of dedicated volunteers.

The tempo of the approaching bicentenary began to quicken. Because of the public visibility of the Engine project, the Science Museum had become both the focus and the clearing house for many other Babbage bicentennial events. Several celebratory conferences were planned, and a Royal Mail special issue Babbage stamp was being prepared. I also still had the Babbage bicentennial exhibition to curate, and there was a seemingly insatiable demand for lectures and articles on Babbage and his work. But one project had priority over everything else: the progress of the Engine and the need to meet the bicentennial deadline – a date that had been sculpted into the cranial folds of our brains – 26 December 1991, Babbage's 200th birthday.

Suppliers delivered parts in batches, and Crick and Holloway began to assemble the machine. We had basic hand tools in the cattle pen. Filing, drilling and reaming were carried out in public view on two workbenches fitted with vices placed behind the Engine inside the 'pen'. Parts requiring machining were taken to workshops in the Museum's basement or sent back to the contracted suppliers.

We did not set out to assemble the machine in its entirety and then test it in one crucial episode. This may have been dramatic, but doing so would run against proven engineering practice. The machine was, after all, a prototype. There were far too many unknowns, and the interaction between the 4,000 parts would make it difficult if not impossible to find faults, which there were bound to be. Instead, the machine was assembled in small stages, and each stage was tested before proceeding to the next.

It was tentative, exploratory work to begin with, as we

developed a feel for Babbage's style. Crick and Holloway first assembled the cast frame which supported all the working parts. The heavy black castings defined the overall shape as a blank rectangle, some eight feet long and seven feet high, that would slowly be filled with intricate, brightly machined parts. Next they started at the end with the crank handle and assembled the large stack of strangely shaped plates, the cams, which generate and orchestrate the complex sliding, lifting and turning motions in an exact and unvarying rhythm. The stack of twenty-eight cams is the heart of the control section of the engine and the core of its 'microprogram'.

In late January 1991 we were ready to try the first experimental turn of the handle with the mechanism only partially assembled. Tentatively at first, and then more forcefully, we applied ourselves to the handle. It would not budge. Even with the special reduction gear which made the machine four times easier to turn, the Engine was locked solid. Ball had warned in 1985 that the engine could not be turned even by a well-motivated operator. Was he right after all? Perhaps Babbage, like his latter-day disciples, had been blinded by optimism, or was too immersed in the intricacies of the machine to spot anything so obvious.

We set about analysing the reason for the resistance. It turned out that the suddenness with which the locking devices needed to accelerate in and out presented a shock load like a hammer blow to the drive handle which the operator was unable to overcome. There was no way to make the motion gentler: the timing cycle was already too crowded, and the short window during which the locks needed to dart in and out could not be extended. The sceptical engineers had been right after all in their misgivings – but for totally unforeseen reasons.

Rather than invent a solution from scratch, if this were

indeed possible, we sought some hint that Babbage might already have had a remedy in his repertoire. He had, and there was no need to look further than the top of the Engine's design for the clue. The largest load on the drive handle is not the frictional resistance of the hundreds of gears, but the force required to lift and lower the weight of the columns of wheels and their long vertical shafts. To relieve the load Babbage provided strong springs at the top of each vertical shaft which in compression support the shafts in their rest position. Instead of the operator having to lift the full dead weight of the shafts and wheels, all that is required is to nudge them up or down from their equilibrium position. Using springs in this way, Babbage had solved the problem of excessive loads, and with this technique as a model Crick devised an ingenious mechanism that used a spring to counterbalance the load on the drive, just as Babbage had done. The device was quickly built and slung unobtrusively under the Engine frame. It worked a treat.

The handle turned, stiffly at first. As the locks darted in and out we heard for the first time the loud rhythmic slapping sound which is the most audible signature of the machine in operation. At least now there was a prospect of driving the great beast.

As the weeks wore on the Engine slowly took shape. The blank area in the frame began to fill with gleaming columns of bronze wheels contrasting dramatically with the muted grey of machined steel. Each day we dreaded a discovery of some oversight that would either invalidate our understanding of the machine in some unrecoverable way or put the completion date out of reach. The nearer to completion we got, the greater the dread. The wealth of problems that arose were solved one at a time by adjustments, tweakings, fitting, trimming and some re-machining, painstakingly carried out by Crick and Holloway in their floodlit engineering zoo.

The visiting public were drawn as though by magnets to the site of the build. Many visited again and again to follow the progress of the machine as it grew. The Engine was like nothing they had ever seen. It was manifestly a Victorian object, yet it resembled none of the machines popularly associated with the nineteenth century, bearing no similarity to a clock, a steam locomotive or a textile machine. The intricate shapes of many of the parts led to musings about how they were made. Questions visitors asked the engineers revealed that they had been drawn to the machine not particularly because of any pre-existing interest in computing or calculation, but because of curiosity about manufacture stimulated by the spectacle of this unique device.

Our early fears about an inquisitive public being an irritation proved unfounded. The questioning visitors gave gratifying feedback that the work was valued, an unusual experience in engineering design and construction, so much of which is a solitary or invisible process. The interest was flattering. There was a freshness to it which renewed the engineers' sense of the value of what they were doing, even when the explanations were familiar and had to be repeated time and time again. Word spread. More visitors came. A French student was overheard in the South Kensington underground station being asked what he was doing in London. He was on his way to the Science Museum to see the Babbage Engine.

Not all went well. We were breaking new ground in taking a Babbage Engine beyond anything Babbage had himself accomplished. There were incessant jams which were maddening – the handle would turn barely a few degrees before the mechanism would lock up. The jams were difficult to deal with. Babbage had not had the experience of fault-finding on a complete machine, and had made no provision for easing the setting-up process or dealing with malfunctions. In modern computer systems and electronic equipment there is provision for fault-finding and correction, a process known in the computer industry as 'debugging'. (The original computer 'bug' was in fact a moth that had lodged in the electrical contacts of an early vacuum tube computer and caused an elusive malfunction. Since then 'debugging' is the name given to the process of finding and fixing faults in both software programs and electronic hardware.)

Modern electronic hardware has facilities to help diagnose faults. The internal states of electronic equipment are often monitored and displayed on indicator lights brought out onto a console. The lights give warnings of faults or malfunctions which are ordinarily hidden from the operator or engineer. Another technique is to bring out test points from inaccessible internal circuitry to a front panel so that the condition of the device can be checked by connecting up external equipment. Software programs are written in self-contained modules to allow piecemeal testing.

Babbage had made no provision for debugging. There is no easy way of isolating one section of the Engine from another so as to localise the source of a jam. The whole machine is one monolithic 'hard wired' unit. Drive rods and links are pinned or riveted permanently into position, and are difficult to dismantle once assembled. All a hapless engineer can do is to poke around with a screwdriver, prising here and there in the hope of finding some play in a link or drive rod. If there is movement, the jam is probably earlier in the machine, so you repeat the process until a moving part is found with no play at all. This is then inspected as the likely site of the jam. It was erratic and frustrating work, and the Engine was highly inventive in producing ever more obscure reasons for seizing up.

When the whole machine locked solid, finding a starting point for fault detection was a haphazard and demoralising business.

The early jams were caused by fouling between parts where the clearances were critical and sharp edges had not been removed by bevelling or chamfering to ease engagement. This was readily remedied by re-machining hundreds of similar parts in the same way or rounding them off by hand with a file if the parts were few. But the most chronic source of jams was not so easily remedied – synchronising the motions. The internal timing and phasing of the complex motions is critical, and if a part moves even fractionally early or late the mechanism is likely to seize or break as it clashes with a mating part.

When parts were first offered up to the machine they were rarely in the correct orientation, and had to be adjusted by trial and error so that their motions harmonised with those of the rest of the mechanism. This is a tentative and exploratory process, and Babbage gives no clues to how the parts are to be located correctly. He blithely shows gears and levers fitted in their correct positions and fixed permanently on their shafts by pins driven through both parts locking them to each other immovably. There is no indication of how the correct rotational position of the gear is to be found before it is finally fixed. So the timing of hundreds of parts had to be determined by meticulous trial and error. In order that parts could be adjusted before pinning, grub screws were used to temporarily fix rotating parts to their shafts. Each part was adjusted incrementally for each trial. As soon as the correct timing was achieved the part was pinned permanently in position and the next part tried. But until all the parts were correctly positioned, the machine would jam and jam and jam again. Each time a small experimental adjustment, and another trial turn. With thousands of moving parts, it seemed as though the sources of jams and breakages was endless. But bit by bit the assembly progressed.

Pinning a part involved drilling through it into the shaft and then driving a short metal rod through both parts by tapping the rod with a hammer – a process that had to be repeated hundreds of times. The drilling was done with a very un-Victorian electric drill. A problem with the process was that swarf and debris from the drilling would fall into the rest of the machine and foul the works. So while Crick drilled away, Holloway stood on a ladder holding an equally un-Victorian Hoover hose over Crick's shoulder to vacuum away the swarf. Twentieth-century electric tools being used on a nineteenth-century machine looked like a ridiculous anachronism. Mentally I apologised to Babbage, and hoped that he understood the needs of the moment.

While Crick and Holloway plugged away at the assembly, the pressures of the bicentenary exhibition began to tell. Apart from unveiling the first working Babbage Engine, the ambition of the exhibition was to display the most comprehensive collection of Babbage-related artefacts ever assembled in one place. Family memorabilia were generously made available by the Babbage family, including a pair of miniature watercolours of Babbage and Georgiana painted during their engagement in 1813. These were loaned by Neville Babbage, great-great-grandson of Charles, who personally delivered the tiny images from Australia. The wooden-framed Scheutz prototype, the first printing calculator completed by Georg and Edvard in 1843, and a compact difference engine devised by Martin Wiberg in the early 1860s were coming from Stockholm. And a splendidly elaborate calculator by Johann Müller dating from 1784 was being transported from Darmstadt. The exhibition would celebrate Babbage's life and work and tell the tale of how, despite his and others' inspired efforts, the movement to mechanise and automate calculation in the nineteenth century had failed. I now had another set of deadlines: authoring the text and identifying illustrations for the exhibition, as well as for an illustrated book to accompany the exhibition.

Staff rallied round. Wendy Sheridan, the Science Museum's curator of the pictorial collection, took on the portrait research. Peter Davison, a staff designer, tackled the panel layouts, typography and illustrations. Gilly Burton, the exhibition designer, immersed herself in the period feel, and created an exhibition design that captured the sepia and mahogany heaviness of Victorian elegance with great subtlety. Neil Brown, responsible for delivering the exhibition, took on the laborious job of organising the foreign loans and the countless tasks associated with contractors, site management, and object movement, insurance and indemnity. Dave Exton, the staff photographer who had produced such startling images of the trial piece, set about photographing sixty objects from scratch to produce colour plates for the book illustrations. Sheridan, sensing the pressure, wordlessly volunteered her time to do the copy-editing.

We worked as an extraordinary ad hoc team. I would draft text overnight and fax it through to Davison and Sheridan the next morning. Davison would do trial layouts while Sheridan edited and proof-read the copy, turning it around in little more than an hour. By early afternoon Davison had corrected copy for faxing to the typesetters. Camera-ready copy would be couriered back for pasting up by evening. We worked this frantic rota for weeks. It was wartime. The enemy was the clock. And throughout it all, daily progress on the Engine was being carefully monitored.

The rewards from the Engine were slow and hard-won. Despite the pressure and uncertainty we could still marvel at the mechanism we were wrestling with. In early March we were rewarded with our first sight of the helical geometry of the carry mechanism. These exquisite assemblies resemble elegant elongated circular staircases. Seven helices rotated in unison, like dancing DNA. The pinpointed reflections of light which caught the mechanism as it turned created the illusion of a rippling, wave-like motion travelling up the machine. It was an arresting, beautiful sight.

Babbage still had the capacity to surprise. The original drawings show horn-like attachments on parts of the carry mechanism whose function had not been fully evident to us. Respectful of Babbage's intentions, we had adhered to the original design and carried these strange horns through to the new drawings. It was only when the device was seen in live operation that their purpose was revealed to be part of a security mechanism that prevents figure wheels from moving unless driven from the correct source. The horns stop the figure wheels deranging even while waiting for a possible tens carry from below. We could only wonder at the ingenuity of such a safeguard, and be humbled by the fact that the old master still had the capacity to astound.

The Engine surrendered its secrets reluctantly. We felt as though it was testing us. Sometimes it teased. Sometimes it sulked. Occasionally it gave us an unexpected reward.

Chapter 17

THE 'IRASCIBLE GENIUS' REDEEMED

The labours of Mr. Babbage . . . are a marvel of mechanical ingenuity and resource.

C. W. Merrifield, 1878

Each day that passed without some awful discovery to scupper the venture allowed us to hope a little more that we might succeed. So far we had found nothing that led us to believe that the Engine was flawed in its conception or its realisation. Countless ordinary calamities were solved on the bench, and so far nothing catastrophic to the larger ambitions of the project had befallen us.

The Engine was to be the centrepiece of the bicentennial exhibition that would include all the significant surviving relics of Babbage's efforts to build his machines. Because of its size, the new Engine would dominate the exhibition space. It was already a substantial addition to the collection of physical artefacts, and the history of its construction represented a modern chapter in the long chronicle of Babbage's thwarted efforts. The gallery booklet that was to accompany the exhibition would include new colour plates of most of the objects on public display. Not to have a photograph of the new Engine in its completely assembled form would be a glaring omission. For the purposes of the photograph it didn't much matter whether or not the Engine worked, but all the parts would have to be in place if the image was to convey a realistic visual impression of the machine. The production schedule for the book required the photograph eight weeks before the exhibition opening. The problem was that the Engine was still incomplete, and the method of testing one small section at a time meant that it would probably remain incomplete until the very end. In May 1991, eight weeks before the exhibition opening, I succumbed to the entirely extraneous need for a catalogue photograph. Knowing that what we did was ill-advised, we abandoned engineering best practice and hastily assembled the full machine without any systematic checking of its operation.

With the photo session complete, the book had secured a handsome prize – the first image of the assembled machine. But the untested sections of the Engine played havoc with the testing process. The machine jammed incessantly, and there was constant fouling and interference from the mechanisms that were being driven without having been correctly phased. With the exhibition opening seven weeks away we were faced with a choice: hack on with the bad-tempered Engine fully assembled, or take the time to dismantle the machine and restart where we had left off before the shoot. We took a gamble and carried on, fighting the uncoordinated jams and crashes from the untested sections of the protesting machine.

During this period we discovered the 'ping' test. If the carry mechanism fouled, the bronze levers, which were brittle, simply snapped. There would be an audible 'ping' followed by a clatter in a far corner as the fractured limb landed. It was like a cricket match with a mad batsman walloping bits of bronze to the boundaries. Someone would forage for the severed part and then do a visual search for damage on the machine. At over £100 a part and a half-day to replace the fractured bit, the 'ping' was something to be dreaded. We turned the handle with a new wariness, and everyone stood poised to duck.

Two weeks before the exhibition opening the machine was moved from the build site to its display position in the exhibition area thirty yards away. It was still jamming, and the progress of the commissioning process was halting. Special stays and struts to brace the engine against distortion in transit were fixed in place, and the Engine glided across the floor on poles to take its new place. Crick and Holloway no longer had the protection of the cattle pen. They were now working on the Engine on a building site as the exhibition went up around them.

At the same time a press notice was widely circulated to newspapers, computer journals, and radio and television companies. The notice announced the exhibition and enclosed an invitation to attend the press view at 9.00 a.m. on Thursday 27 June. There was no going back. While the notice did not explicitly say that the Engine worked, it did nothing to discourage the media from believing that they would be privileged to witness the first public display of a working Babbage Engine, and be among the first to behold a sumptuous spectacle that no Victorian ever saw.

Science Museum
Press Notice
Embargoed until 12.00 noon, Thursday 27 June 1991

MAKING THE DIFFERENCE AT THE SCIENCE MUSEUM
CELEBRATING THE BICENTENARY OF THE BIRTH
OF CHARLES BABBAGE

A special exhibition, 'Making the Difference – Charles Babbage and the Birth of the Computer', celebrating the bicentenary of the birth of Charles Babbage, opens to the public at the Science Museum tomorrow, Friday, 28 June 1991.

Charles Babbage (1791–1871), the English mathematician, is the towering ancestral figure of the history of computing. He has been dubbed the 'irascible genius' and also been called 'one of the noblest and most ingenious sons of England'. He invented automatic calculating machines over a century before the electronic era but failed to build them.

The centrepiece of the exhibition is Babbage's Difference Engine No. 2, the first full-size Babbage engine ever completed. The engine has been built by the Science Museum to Babbage's original designs dating from 1847 to 1849. By building this engine the Museum set out to prove that these machines could have been built in Babbage's day and so redeem the reputation of the first computer pioneer in time for the 200th anniversary of his birth.

The tempo in the gallery was rising as preparations for the opening and the press view drew near. Fitters, designers, contractors, lighting engineers, carpet layers, case dressers, window cleaners, managers, press officers and curators buzzed around the area. We had still not managed one complete cycle of the Engine without a jam, but unless we hit an insurmountable obstacle we still felt we were in with a chance. Crick and Holloway doggedly kept to their programme of systematically adjusting, testing and pinning in the midst of the babble of preparations. The Engine looked magnificent on its mahogany plinth. The black cast frames contrasted with the steels and bronzes, and the bright finish more than justified the extra cost of machining from solid instead of casting. But still it would not work. It was as though the machine was teasing us with its good looks as it hid its secrets.

Gradually the jams became less frequent and I began to think ahead about the first calculation we would try. I went off home one night and programmed an old personal computer (a North Star Horizon I had bought in the United States in 1979) and, using an early version of Multiplan, a spreadsheet program, produced a printout of the first eighty values of the first calculation we would perform: a table of the counting numbers (1, 2, 3, etc.) each multiplied by itself seven times. This calculation, a table of powers of seven, would fully exercise the Engine, and the results read off from the number wheels could be compared with the results calculated by the electronic computer. I also produced a table of the initial values that the number wheels would need to be set to at the start of the first test run.

On 24 June, three days before the public opening, we attempted to set up the first calculation using the procedure described by Babbage in a rare breach of his usual practice of putting nothing in writing to explain the workings of his machines. This procedure involved advancing the Engine to a predetermined point in the cycle, disengaging the security locks, and turning by hand the number wheels on each alternate column to enter the numbers calculated in the table prepared beforehand. The Engine would then be advanced to the next point in the cycle, and the process repeated for the remaining four columns. This is a one-off operation. Once completed, the machine would produce each next value in the table without further intervention – if, that is, the thing worked at all.

We carefully followed Babbage's procedure. We advanced the Engine to the first setting-up point, disabled the security locks, and warily rotated the wheels by hand to register the values on the preprinted sheets. So far, so good. As we advanced the Engine to the second setting-up position we found the oversight we had dreaded for so long: advancing the Engine deranged the wheels we had adjusted only moments before. Babbage had omitted to provide a means of locking the wheels during the procedure, and the setting-up sequence was self-corrupting. The solution was simple enough. We would need to provide temporary locks – metal strips inserted into the gears – to hold the settings on the first four columns while we set the remaining four. But we had almost no time to do this. We cadged strips of sheet metal from the workshops and clipped them onto the Engine frame with ordinary bulldog clips filched from someone's stationery drawer. The makeshift locks worked. Crisis was averted – at least for the time being.

With fewer jams and a trial calculation in prospect, we dared to hope that we might just get in under the wire in time for a triumphant announcement at the exhibition opening. Then, on 25 June, with two days to go, as one of the engineers gingerly turned the handle, a defective weld in the cam stack failed and a critical part, shaped like a wishbone, snapped in two at the join.

To recover the part we would have to dismantle the cam stack. This would take two days, overrunning into the opening. It would take a further day to repair the broken part and another two to reassemble the stack and get us back to where we were. The contractual safeguards obliging the suppliers to remake defective parts at their own expense now meant nothing. Time was the critical commodity. Were it not for this failure, we still believed we had a chance to get the machine to work in time. There was a way out, but it was a drastic one – to cut through the pillars of the frame to retrieve the part and make good the damage by means of a new pillar that could be assembled with the rest of the frame intact. Holloway took a hacksaw and, in an act that in other circumstances would rank as one of gross vandalism, sawed through the pillar. A crash team was waiting in the workshops. They rewelded the broken part, and had it filed and painted and ready for reinstallation within hours. While the emergency repair was in progress another technician turned a new pillar on a waiting lathe. We had lost a day. There was one day to go.

With the repair complete, four of the eight columns seemed to be working correctly. But the remaining four columns, the ones we had hastily assembled for the catalogue photograph, kept jamming. We tried disabling all the carry mechanisms in the hope of eliminating the problem, and ran the calculation without carries to get the machine to cycle freely. But the Engine would not be hurried, and jammed again.

Television and press reporters gather in the early-morning sunshine outside the Science Museum waiting to be let in. Inside, senior Museum staff assemble in the director's suite for a briefing before the press view. Neil Cossons, marketing managers, press officers and divisional heads stand in a loose circle. The press officer runs through the list of TV channels, newspapers and scientific journals that will be covering the public unveiling of the new Engine. On the cover of the latest issue of *New Scientist*, out that day, a grumpy Babbage sits with a laptop on his knees, a personal computer in the background and a portion of the new Difference Engine clearly in view. 'Babbage: Architect of Modern Computing', it announces.

Inside is a long feature article on the Engine project. There will be a supply of complimentary copies available as handouts to satisfy media interest in the person they have regarded, rightly or wrongly, as the progenitor of the modern computer age.

One of the press officers summarises the presentation schedule. Cossons will open with a brief address, and I will follow to talk about the Engine. We will then take questions and finally demonstrate the Engine by performing a calculation. I make a coughing sound. I explain that while we have good reason to believe that the machine will eventually work, we have been unable to solve the jamming in time, and that attempting a full calculation would entail an unacceptable and possibly catastrophic level of risk. Instead we would cycle the machine with the number wheels all set and locked at zero. The Engine would make the loud rhythmic clanking sound characteristic of its operation, and the helical carry mechanism would be fully visible. The look and sound of the engine calculating with zeros would be little different from its behaviour during a full-blown calculation. We had no option. It would have to do.

The statement is met with visible confusion, scarcely concealed anger and mumbled accusations of blame. A ragged debate follows on how we are to play the media's expectation that they have come to witness a fully working engine and the first public calculation. Time ticks by. The press view is due to start. Nothing is resolved. We file out of the director's suite and troop through the great oak doors down to the public gallery on the ground floor. The hubbub from the assembled press gets louder as we approach.

The Engine stands resplendent on its polished mahogany plinth. The track lights trained on the new machine bathe it in a pool of yellow light. Surrounding it, like broken promises, are the relics of unsuccessful attempts from the nineteenth century, by Babbage, Scheutz and others. The new Engine towers above the gathering. The relics in their glass display cases are dwarfed by the new incumbent, like courtiers to the grand new Engine which stands as the final fulfilment of all those early dreams and bleak struggles.

The gallery is packed. The press has assembled in the apron area in front of the great machine. There are TV cameras and blazing lights on tripods like pylons in a landscape of bobbing heads. Nikons, Canons and Leicas with flash guns on squat stalks are at the ready. The stage is set.

Cossons and I step onto the mahogany plinth with the great Engine as a backdrop. We face the lights and the sea of upturned faces. I have no idea what we are going to say.

The gallery goes quiet. Cossons explains that the exhibition is to commemorate the work of the first computer pioneer and draws attention to the presence of the most authoritative and comprehensive collection of Babbage-related material ever gathered in one place. He pays tribute to the Science Museum team which has undertaken the grand venture to accomplish what Babbage had failed to do. He says that he understands the avid interest in the new Engine and what it signifies. With a final welcome he hands over to me.

There is a rustle of movement. Eyes and cameras pan the short distance to the figure alongside him. I am candid about where we have succeeded and where we have failed. I emphasise that we are very close to achieving a complete working Engine and that we have found nothing that weakened our belief that the Engine would work. I invite questions. To our relief it is clear that the ambitions of the project and the appearance of the strange machine have captured the reporters' interest sufficiently to deflect attention away from the overall inconclusiveness of the historical programme. But there is some uncomfortable barracking about what exactly it is we think we have proved. And they are impatient to see the Engine work.

Crick turns the handle. The rhythmic clanking and the shifting array of bronze wheels begins. The helical carry mechanisms perform their rippling dance. There are murmurs as the motions enthral and seduce. The visual spectacle of the engine works its magic. As a static exhibit the engine is a superb piece of engineering sculpture. As a working machine, even partially working, it is arresting. Cameras caress the Engine and dwell on its intricate moving parts. Any anxiety we had that the media might see the Engine as an elaborate failure in the grand tradition of all the earlier failures is quickly dispelled. The Engine has cast its spell, and later that day the coverage is ungrudgingly triumphalist. But to the tutored eyes of the team the unmoving zeros on all the number wheels are a private signal that our work is not yet done.

The exhibition opened to the public the next day. Visitors flocked to see the new machine. In response to public interest fuelled by television, newspaper and radio coverage, Crick and Holloway demonstrated the Engine twice daily, in the morning and afternoon. The advertised demonstrations took time, interrupted the testing and debugging procedures, and confined any real work to the hours between public performances. These demonstrations also prevented us from undertaking any works that would render the machine inoperable for more than a few hours. Progress was slower than ever. There were days when public interest was such that no testing was done at all.

I knew that the press view represented an unrepeatable opportunity for a grand coup. Privately we were disappointed that we had not made the deadline, but we also knew that while coverage was politically crucial, this was quite separate from the serious purpose of getting the Engine to work and, if at all possible, to do so by 26 December, Babbage's 200th birthday. Despite the bumpy ride so far, we had not yet really found anything that compromised that prospect.

In the months that followed the engineers worked during the short intervals available. The jams became fewer and then became rare. With use, the initial stiffness began to ease. We tried the prepared test calculation, powers of seven ($y = x^7$) for the first 100 values, and noticed a strange phenomenon: phantom carries. Numbers appeared in sections of the Engine where no calculation was taking place – a random sprinkling of '1's appearing in the upper sections of the column where there should have been zeros. The columns were removed and stripped. We found that the tolerance on one of the parts – a part repeated many times over – was out of specification. The hundreds of similar parts had to be removed from the engine, trimmed and reassembled. The phantoms were banished.

There was no sudden dramatic point at which a crucial all-or-nothing test succeeded. The slow path to reliable working was a gradual process of eliminating more and more refined forms of minor fault. In late November we were nearly home and dry, with only a few erratic carries that failed to operate. We had thought that with modern production techniques the repeat parts would be so nearly identical as to be interchangeable, but we found otherwise. Babbage had insisted on the highest precision in the construction of his Difference Engine No. 1, and the latter-day team found that his insistence was well founded. About twenty of the 210 carry mechanisms needed hand-tuning, and Crick made a strangely shaped tool to tweak the out-of-true limbs. Each time a carry mechanism faltered a lever was bent a little by hand and retested. On Friday 29 November 1991 the engine performed its first full automatic error-free test calculation. The calculation was repeated time and time again. The machine was faultless. We had done it. We had built the first Babbage Engine, complete and working perfectly, twenty-seven days before Babbage's 200th birthday. All we had to do now was build the printing apparatus. But that is another tale for another day.

Chapter 18

THE MODERN LEGACY

Another age must be the judge.

Charles Babbage, 1837

Charles Babbage is routinely referred to as the father, grandfather, forefather, great ancestor or progenitor of the modern computer. The image of fatherhood reinforces the notion of an unbroken line of descent with Babbage as the patrilinear source. Because Babbage was the first to embody in his designs the principles of a general-purpose computing engine, it is often assumed that the modern computer has descended directly from his work. But the lineage of the modern computer is not as clear-cut as these genealogical tributes imply.

History has transformed Babbage from a colourful figure in nineteenth-century science into a national hero. In 1991, the bicentennial year of his birth, major computer companies sponsored a commemorative exhibition at the Science Museum and, by implication, appropriated him as the ancestral patron of the industry. The Royal Mail launched a special issue of four postage stamps commemorating British scientific achievement. Babbage shared the honours in the philatelic hall of fame alongside Michael Faraday (electricity), Frank Whittle (jet engine) and Robert Watson-Watt (radar). The implication is clear – that Babbage contributed as much to modern computing as Faraday (whose bicentenary he shares) did to our knowledge of electricity, or Whittle and Watson-Watt to our understanding of the jet engine or radar. Of the four, as we shall see, Babbage is the odd man out.

Scientific American, *New Scientist* and the *Bulletin of the British Computer Society* ran feature articles on the 'architect of modern computing'. The received wisdom is captured by a statement in the issue of *Scientific American* which featured a specially commissioned image of Difference Engine No. 1 on the cover. The caption reads simply, 'Charles Babbage's plans for mechanical calculators and computers paved the way for the modern computer revolution.' But Babbage's influence on modern computing may not be as strong as popular perception would have us believe.

In 1991 Allan Bromley, computer scientist and historian, who had studied Babbage's designs more closely than anyone, wrote that 'Babbage had effectively no influence on the design of the modern digital computer.' Maurice Wilkes, distinguished pioneer of the electronic computer, had come to the same conclusion. In 1971, the centenary of Babbage's death, Wilkes published an article entitled 'Babbage as a computer pioneer' in which he wrote that Babbage, 'however brilliant and original, was without influence on the modern development of computing'.

In the same article Wilkes, who elsewhere describes Babbage as possessing 'vision verging on genius', goes further than claiming that Babbage did not pioneer the modern computer age, actually accusing him of delaying it. Wilkes argues that Babbage's projected image became one of failure, and that this discouraged others from thinking along similar lines. I first read Wilkes's claim as a rhetorical device intended to counteract unexamined glorification of Babbage – an attempt, as it were, to bend the celebratory wire beyond its neutral point in the opposite direction to offset any tendency to hysterical excess. But new evidence has come to light of at least one instance for which Wilkes's allegation, however originally intended, is specifically and historically valid.

Thomas Fowler, a self-taught Devonshire printer and bookseller, devised an original computing device. Fowler's machine differed in essential respects from Babbage's. Instead of using gearwheels to represent the numbers '0' through '9' by ten discrete positions of the wheel, as Babbage had done, Fowler's machine used sliding rods, each of which could occupy only one of three positions to validly represent the digit of a number. The advantage of having fewer distinct physical states is that it is easier to discriminate between them, and parts can be made less precisely without the device malfunctioning. Fowler was too poor to work in metal, and made working versions of his calculator in wood in 1840 and 1842. His digital calculators were demonstrated at the Royal Society, the British Association, and exhibited at King's College to some acclaim. The machine calculated logarithms to thirteen places 'in a singularly beautiful and concise manner' and the principle could be adapted from ternary to binary working. His calculator was a scientific novelty, and luminaries of the mathematical establishment flocked to view it. Babbage, the astronomer Francis Baily and the logician Augustus De Morgan saw the machine in action and were impressed by its originality, speed and efficiency, even though, unlike Babbage's machine, it was not completely automatic.

In a biographical notice published in 1875, Fowler's son

wrote of his father's struggles to realise his machines. The refrain will be familiar from Babbage's story, but these are Hugh Fowler's words musing on his late father's efforts:

> It is sad to think of the weary days and nights, of the labour of hand and brain, bestowed on this arduous work, the result of which, from adverse circumstances, was loss of money, loss of health, and final disappointment.

Thomas Fowler had neither social position nor means. He was far from the bustle and self-confidence of the London set and worked in almost complete isolation. With great humility he hoped to exhibit his machine at a forthcoming meeting of the British Association for the Advancement of Science, and wrote in 1841 with touching vulnerability to the Astronomer Royal, George Biddell Airy:

> I have led a very retired life in this town without the advantage of any hints or assistance from anyone, and I should be lost amidst the crowd of learned and distinguished persons assembled at the meeting, without some kind friend to take me by the hand and protect me.

Fowler was paranoid about his invention being stolen. His fears were well founded. In 1828 he had invented and patented the Thermosiphon – a device for circulating hot water through pipes. Despite the patent, the invention was pirated and widely used, and Fowler was robbed of all benefit. After this he refused to release drawings or diagrams of his engine, and none have since been found though some written descriptions survive. Hugh Fowler recalls the painful memory of one description 'having been dictated by my father to my sister on his death-bed, while in great suffering from the disease of which he soon after died'. It is with unmistakable bitterness that Fowler's son writes:

> The government of the day refused even to look at my father's machine, on the express ground that they had spent such large sums, with no satisfactory result, on Babbage's 'calculating engine'.

With the benefit of hindsight, we can see that Fowler's machine is in certain respects vastly more promising than Babbage's. The use of sliding rods or plates reappeared in the late 1930s and early 1940s when the German pioneer of computing Konrad Zuse built automatic programmable calculators using the technique in his binary (two-state) digital machines. Zuse appears to have had no knowledge of Babbage or of Fowler. In the light of Zuse's work and the near-universal adoption of binary for electronic computers, Fowler's machines can be seen to be closer to modern digital circuitry than Babbage's.

As a historical figure there are respects in which Thomas Fowler in the nineteenth century is to computing as John Harrison is to clocks in the eighteenth. Fowler devoted a large part of his efforts to devices for the production of rapid error-free calculation; Harrison to accurate time pieces for the determination of longitude. Neither was wealthy, both were self-taught, completely original and worked almost entirely alone. As with Harrison, the promise of Fowler's work was not explored by his contemporaries. In Fowler's case this seems to have been directly as a result of Babbage's failed efforts.

The Fowler episode was unknown to Wilkes in 1971 (I found the Fowler correspondence in 1993), and a few years ago I asked him what he had in mind when he alleged that Babbage had delayed the development of computers. Wilkes was quite clear in his reply. He had known L. J. Comrie, an acknowledged authority on the calculation and production of mathematical tables. Comrie was Superintendent of the *Nautical Almanac* from 1930 to 1936, and spent eleven years there involved in the annual publication of printed tables for navigation. He knew of Babbage's earlier attempts, and the lesson he drew was not to undertake the construction of purpose-built machines but to wait for commercially developed machines which could then be adapted. Comrie used a National Accounting desktop calculating machine and later a Burroughs machine for detecting errors in existing tables by repeated subtraction (as suggested by Airy), as well as for the generation of new tables using repeated addition, as in Babbage's schemes. With Comrie's work, difference engines, finally, were being used routinely for automatic tabulation. Comrie is reported to have remarked that 'this dark age in computing machinery, that lasted 100 years, was due to the colossal failure of Charles Babbage'. Wilkes had meant by his remark that Babbage's experiences had discouraged the development of purpose-built automatic calculating machines. Who is to say what might have happened were it not for the traumatic precedent of Babbage's tale?

Babbage, the Scheutzes and Fowler were not the only ones to attempt automatic calculating machines in the nineteenth century. Some succeeded, but these were isolated sputterings all of which came to nothing. Alfred Deacon in London, stimulated by Lardner's description of Babbage's work, built a small difference engine for his own amusement. Martin Wiberg in Sweden built two compact desktop differencing calculators, the first of which was finished in about 1860, which he used to print and publish volumes of tables in Swedish, German,

French and English. But the machine failed commercially, and the typography was inelegant. Barnard Grant in America built a monster machine – over a ton in weight, eight feet long and five feet high, with 15,000 parts. It was exhibited at the Philadelphia Centennial Exhibition in 1876. It vanished without trace. In terms of hardware these initiatives led nowhere. In terms of technique it was not until Comrie's work in the twentieth century that tabulation using mechanical differencing was used to any great extent.

The difference engines attempted in the nineteenth century were automatic calculators, not general-purpose programmable computing machines like Babbage's Analytical Engine. With Babbage's death, automatic general-purpose computing died too. There was one febrile twitch in the first decade of the twentieth century. Percy Ludgate, an Irish auditor, designed an 'analytical machine' between 1903 and 1909. The machine is an electrically driven programmable calculator using perforated tape to input instructions. The design is original, and Ludgate attested that he had no prior knowledge of Babbage's work. Ludgate's machine is an isolated episode, a developmental cul de sac, with no discernible influence on what followed.

It is clear from this that there is no continuous line of technological development from Babbage to the electronic era. But the gulf between the two is not total. There were those who kept his name alive. One of these bridging figures is none other than Babbage's son Henry, whom we left at his father's bedside.

Henry inherited the parts, the drawings, the experimental work and all the effects relating to his father's Engines. Babbage's workshop included forges, fly presses, several lathes, a massive planing machine, a high-pressure steam engine, a grooving machine, drilling and slotting machines, benches, vices, hand tools, and a mass of fittings and scrap. In March

1872, while on furlough from service in India, Henry auctioned the contents of the workshop but retained a few lathes, a small planing machine and some tools. During the two years of his furlough he continued diligently with his father's work, and then intermittently, after he returned to England in retirement in 1875. He retained one of the workmen, and in 1872 assembled the small experimental section of the Mill of the Analytical Engine under construction at the time of his father's death. The Mill uses pressure-diecast parts, and includes sections of the anticipating carriage mechanism, some columns of the Store and horizontal toothed racks which transfer information between the sub-units. The machine is incomplete and it is unlikely to have functioned, even as a demonstration piece. It was assembled as a memorial – a tribute from an admiring son – and is now on display in the Science Museum in London. Its small size gives little indication of the monumental intellectual achievement its conception represents. It survives as a marker, a token of what was not to be.

Henry felt that the bequest from his father was a demonstration of trust, and he wished to repay this confidence 'by embodying some of his ideas in metal'. He built an arithmetical section of the Mill of the Analytical Engine – a four-function calculator. He had the bed cast in 1888 and the machine, fitted with a printing section, was working by 1910. Henry was then eighty-six years old. The machine calculated and printed the first twenty-two multiples of π. But there were errors. It was deposited at the Science Museum and borrowed for various exhibitions, including the Japan-British Exhibition in White City, Shepherd's Bush, in 1910. But there was no great interest in it. Henry's efforts were conscientious but without the inventive inspiration of his father. The world had moved on, and with Herman Hollerith's punched-card equipment stealing the limelight, Henry's machine was a legacy of a vanishing past. After the Coronation Exhibition in 1911 it was returned to the Science Museum in London, where it sits today in solid repose.

The final fate of Clement's work on the Difference Engine is a dismal one. Babbage's bequest to Henry included most of the unused parts made by Clement for the Difference Engine – almost enough to assemble the calculating section of the machine. Most of the framing plates had been cannibalised by Babbage for other work. After his final return from India in 1875 Henry paid to have them remade and started assembling the machine that had been abandoned so abruptly by Clement over forty years earlier. Eventually he gave up the project. About 10,000 parts made to the highest precision of the day were consigned to the melting-pot for scrap. But all was not lost. Henry retained enough of Clement's parts to make five or six small Difference Engine demonstration pieces. These he sent to various universities, including Cambridge, University College (London) and Manchester, to advertise his father's work. He offered one to Harvard University in 1886, and the model sent there provided another tenuous link with the modern computer age.

The movement that led to the modern computer did not resume until the 1940s, when pioneers of the electronic computer laid the foundations for what followed. Although the development of general-purpose computing engines largely stopped with Babbage's death, the question of whether his work none the less influenced the pioneers of the modern computer age is an intriguing one.

The mechanical era of computing was followed in the late nineteenth century by an electromechanical era ushered in by Hollerith's punched-card equipment. Some of the pioneers of electromechanical apparatus were aware of Babbage, others not. The same is true of the early electronic era, except that here awareness was more widespread. But almost without exception the seminal figures disclaim any material influence on their work.

There is one exception. Howard Aiken, who pioneered the construction of the Harvard Mark I in the United States, is one pioneer of the modern era who specifically claimed to have been influenced directly by Babbage's work. The Harvard Mark I, an electrically driven program-controlled calculator, was built in an IBM plant in Endicott, New York, and completed in January 1943. It was a monster – fifty-one feet long, two feet deep, and eight feet tall, weighing five tons and consisting of 750,000 parts. It was a highly publicised and influential machine but was a technological dead end.

In the late 1930s Aiken came across the small demonstration piece that Henry had sent to Harvard. Aiken later claimed that he 'felt that Babbage was addressing him personally from the past' and that 'if Babbage had lived seventy five years later, I would have been out of a job'. Aiken repeatedly emphasised his indebtedness to Babbage, and his frequent tributes publicised Babbage's work in the post-war years. Aiken styled himself as Babbage's modern-day heir. It is curious that the historian I. B. Cohen has gone out of his way to demonstrate not only that Aiken was largely ignorant of the detail of Babbage's work, but that some of his perceptions were in fact wrong. Cohen accuses Aiken of seeking bandwagon fame – of attempting by his public affiliation with Babbage to stake a claim to his own place in history. It is an irony that the one pioneer to lay so strong a claim to direct influence is accused of immodest self-promotion. History, it seems, is determined that Babbage shall have no intellectual heirs.

Babbage published practically nothing in the way of technical description of his Engines, and his drawings, which remain largely unpublished in a manuscript archive, were not studied in any significant detail until the 1970s. It is fairly conclusive, therefore, that his designs were not the blueprint for the modern computer, and that the pioneers of the electronic age reinvented many of the principles explored by Babbage in almost complete ignorance of the detail of his work. Such continuity as there is is not in the technology or in the designs, but in the legend. Babbage and his efforts were an inseparable part of the folklore shared by the small communities of scientists, mathematicians and engineers who throughout remained involved in calculation, tabulation and computation. Babbage's failures were failures of practical accomplishment, not of principle, and the legend of his extraordinary Engines was the vehicle not only for the vision but for the unquestioned trust that a universal automatic machine was possible. The electronic age of computing was informed by the spirit and tradition of Babbage's work rather than by any deep knowledge of his designs, which have attracted detailed attention only in the last few decades.

The prehistory of computing is dominated by the towering figure of Charles Babbage. Towards the end of his life he wrote prophetically of his work. The passage below can be read as an expression of his own recognition that he had identified something fundamental, and even universal, about the principles of machine-like behaviour. It can also be read as a declaration of confidence in the judgement of posterity.

> If, unwarned by my example, any man shall undertake and shall succeed in really constructing an engine . . . upon different principles or by simpler mechanical means, I have no fear of leaving my reputation in his charge, for he alone will be fully able to appreciate the nature of my efforts and the value of their results.

CHARLES BABBAGE: BIOGRAPHICAL NOTE

Charles Babbage was born in Walworth, Surrey, on 26 December 1791. He was the first son born to Benjamin Babbage, a London banker, and Elizabeth Plumleigh Teape. Both parents came from established Devonshire families. Two brothers, one born in 1794, the other in 1796, died in infancy. His sister, born in 1798, outlived him.

He was schooled at first near Exeter in Devon, then at Enfield, Middlesex, then near Cambridge by a clergyman-tutor, then at Totnes Grammar School, and finally in classics by an Oxford tutor. In April 1810 he entered Trinity College, Cambridge, where he studied mathematics. He graduated from Peterhouse in 1814, and received his MA in 1817.

In 1814 he married Georgiana Whitmore in Teignmouth, Devonshire, and had at least eight children, three of whom survived to maturity, Benjamin Herschel, Dugald Bromhead and Henry Prevost. His wife, his father and two of his children (Charles and a newborn son) all died in 1827. In 1834 his only daughter, Georgiana, died while in her teens.

The couple made their home in London at 5 Devonshire Street, off Portland Place, in 1815. In 1828 he moved to 1 Dorset Street, Manchester Square, which remained his home until his death.

Babbage was elected a Fellow of the Royal Society in 1816. He was presented with the first Gold Medal of the Astronomical Society in 1824 for his invention of the calculating engine. He was elected to the Lucasian Chair of Mathematics at Cambridge in 1828, a post he held until 1839. Between 1813 and 1868 he published six full-length monographs and eighty-six scientific and miscellaneous papers.

Charles Babbage died on 18 October 1871, and was buried at Kensal Green Cemetery in London.

BIBLIOGRAPHY

Charles Babbage published six full-length monographs and at least eighty-six papers during his lifetime. His known published works have been gathered in an eleven-volume edition, edited by Martin Campbell-Kelly (William Pickering, London, 1989), and this remains the most accessible single reference source. The set contains the six full-length works, including Babbage's autobiographical *Passages from the Life of a Philosopher* (1864), and five volumes of mathematical, scientific, political, economic and philosophical papers. It also includes fourteen items not written by Babbage, notably the classic paper by Ada Lovelace, *Sketch of the Analytical Engine* (1843) and Dionysus Lardner's *Babbage's calculating engine* (1834). The list of works by Babbage cited below includes his major monographs but is not comprehensive with regard to his many scientific and miscellaneous papers.

Ackroyd, Peter. *Dan Leno and the Limehouse Golem*. London: Minerva, 1994.

Aiken, Howard. 'Historical Introduction' in A *Manual of Operation for the Automatic Sequence Controlled Calculator*, 1–9. Cambridge, Massachusetts: Harvard University Press, 1946.

Airy, George Biddell. 'On a peculiar defect in the eye, and a mode of correcting it'. *Cambridge Philosophical Society*, 21 February 1825.

'On Scheutz's Calculating Machine'. *Philosophical Magazine and Journal of Science* XII, July–December 1856: 225–226.

'Further observations on the state of an eye affected with a peculiar malformation'. *Cambridge Philosophical Society*, 20 March 1870.

Airy, Wilfrid (ed.) *Autobiography of Sir George Biddell Airy*. Cambridge: Cambridge University Press, 1896.

Aspray, William (ed.) *Computing before Computers*. Ames: Iowa State University Press, 1990.

Babbage, Charles. *A letter to Sir Humphry Davy, Bart., President of the Royal Society, on the application of machinery to the purpose of calculating and printing mathematical tables*. London: J. Booth, 1822.

'A note respecting the application of machinery to the calculation of astronomical tables'. *Memoirs of the Astronomical Society* 1 (1822): 309.

A comparative view of the various institutions for the assurance of lives. London: J. Mawman, 1826.

'Notice respecting some errors common to many tables of logarithms'. *Memoirs of the Astronomical Society*, 3 (1827): 65–67.

Table of Logarithms of the natural numbers, from 1 to 108,000. London: J. Mawman, 1827.

Reflections on the Decline of Science in England, and on some of its causes. London: B. Fellowes, 1830.

The ninth Bridgewater treatise: a fragment. London: John Murray, 1837.

On the Economy of Machinery and Manufactures. London: Charles Knight, 1832.

The Exposition of 1851; or, views of the industry, the science and the government of England. London: John Murray, 1851.

Passages from the Life of a Philosopher. London: Longman, 1864.

Babbage, Henry Prevost (ed.). *Babbage's Calculating Engines: a collection of papers by Henry Prevost Babbage*. London: Spon, 1889. Reprinted in *The Charles Babbage Institute Reprint Series for the History of Computing*, Vol. 2. Los Angeles: Tomash, 1982.

Memoirs and Correspondence of Major-General H.P. Babbage. London, 1910.

Babbage, Neville F. 'Autopsy report on the body of Charles Babbage'. *The Medical Journal of Australia*, 154 (1991): 758–9.

Ball, W. W. Rouse. *A History of the Study of Mathematics at Cambridge*. Cambridge: Cambridge University Press, 1889.

Barton, H. 'Charles Babbage and the beginning of die casting'. *Machinery and Production Engineering*, 27 October 1971: 624–31.

'Pressure Diecasting in the Eighteen-sixties: Charles Babbage's use of the technique on his "calculating engine"'. *Diecasting and Metal Moulding*, May/June 1972: 14–15.

Becher, Harvey W. 'Radicals, Whigs and conservatives: the middle and lower classes in the analytical revolution at Cambridge in the age of aristocracy'. *British Journal for the History of Science*, 28 Part 4, no. 99, December 1995: 405–26.

Bowles, Mark D. 'The Idealistic Stimulus of Technology; Charles Babbage and the Drive for Reform'. : Cape Western Reserve University, 1993.

Brock, William H. *Science for All: Studies in the History of Victorian Science and Education*. Variorum Collected Studies Series: Aldershot: Variorum, 1996.

'The selection of the authors of the Bridgewater Treatises'. *Notes and Records of the Royal Society of London* 21 (1966): 162–79.

Bromley, Allan G. 'Charles Babbage's Analytical Engine, 1838'. *Annals of the History of Computing* 4, no. 3 (1982): 196–217. Reprinted: *IEEE Annals in the History of Computing* 20, no. 4 (1998): 29-45.

'The Evolution of Babbage's Calculating Engines'. *Annals of the History of Computing* 9, no. 2 (1987): 113–36.

'Difference and Analytical Engines'. In *Computing before Computers*, ed. William Aspray. Ames: Iowa State University Press, 1990.

The Babbage papers in the Science Museum: a cross-referenced list. London: Science Museum, 1991.

Buxton, Harry Wilmot. *Memoir of the life and labours of the late Charles Babbage Esq. F.R.S.* Vol. 13, *The Charles Babbage Institute Reprint Series for the History of Computing*. Cambridge, Massachusetts: Tomash, 1988.

Campbell-Kelly, Martin. 'Charles Babbage's Table of Logarithms (1827)'. *Annals of the History of Computing* 10, no. 3 (1988): 159–69.

(ed) *The Works of Charles Babbage*. 11 vols. London: William Pickering, 1989.

'Large-scale data processing in the Prudential, 1850-1930'. *Accounting Business and Financial History* 2, no. 2 (1992): 117–139.

(ed) *Charles Babbage: Passages from the Life of a Philosopher*. London: William Pickering, 1994.

'Charles Babbage and the Assurance of Lives'. *IEEE Annals of the History of Computing* 16, no. 3 (1994): 5–14.

'Information Technology and Organizational Change in the British Census, 1801-1911'. *Information Systems Research* 7, no. 1 (1996): 15.

and William Aspray. *Computer: A History of the Information Machine*. New York: Basic Books, 1996.

Ceruzzi, Paul E. *Reckoners: The Pre-history of the Digital Computer*, *from Relays to the Stored Program Concept.* Westport, Connecticut: Greenwood Press, 1983.

Chapman, Allan. 'Private Research and Public Duty: George Biddell Airy and the Search for Neptune'. *Journal for the History of Astronomy* 19, no. 57 (1988): 121–39.

'Science and the Public Good: George Biddell Airy (1801-92) and the Concept of a Scientific Civil Servant'. In *Science, Politics and the Public Good: Essays in Honour of Margaret Gowing*, edited by Nicolaas A. Rupke, 36–62. London: Macmillan, 1988.

'Britain's First Professional Astronomer'. In *Yearbook of Astronomy*, ed. Patrick Moore. London: Sidgwick and Jackson, 1992.

'The Pit and the Pendulum: G.B. Airy and the Determination of Gravity'. *The Antiquarian Horological Society* 21, no. 1 (1993).

Cohen, I. B. 'Babbage and Aiken'. *Annals of the History of Computing* 10, no. 3 (1988).

Howard Aiken: Portrait of a Computer Pioneer. Cambridge, Mass.: MIT Press, 1999.

Collier, Bruce. 'The Little Engines that Could've: The Calculating Machines of Charles Babbage', Thesis (PhD), Harvard University, 1970. Facsimile edition with new introduction. New York: Garland, 1990.

Comrie, L. J. 'Modern Babbage Machines'. *Bulletin of the Office Machinery Users' Association Limited* (1931): 29.

'Computing the Nautical Almanac'. *Nautical Magazine*, July (1933): 33–48.

Cullen, M. J. *The Statistical Movement in Early Victorian Britain: The Foundations of Empirical Social Research.* New York: Harvester Press, 1975.

Davis, Martin. 'Mathematical Logic and the Origin of Modern Computing'. In *The Universal Turing Machine: A Half-Century Survey*, edited by Rolf Herken, 140–74. Oxford: Oxford University Press, 1988.

De Morgan, Augustus. 'Description of a calculating machine,

invented by Mr. Thomas Fowler, of Torrington in Devonshire'. : Royal Society, 1840.

De Vries, Leonard, ed. *Victorian Inventions*. London: John Murray, 1971.

Desmond, Adrian, and James Moore. *Darwin*. London: Michael Joseph, 1991.

Dubbey, J. M. *The Mathematical Work of Charles Babbage*. Cambridge, London, New York, Melbourne: Cambridge University Press, 1978.

Eyler, John M. *Victorian Social Medicine: The Ideas and Methods of William Farr*. Baltimore: Johns Hopkins University Press, 1979.

Farr, William, ed. *English Life Table: Tables of Lifetimes, Annuities and Premiums*. London: HMSO, 1864.

'Scheutz's Calculating Machine and its Use in the Construction of the English Life Table No. 3'. In *English Life Table: Tables of Lifetimes, Annuities and Premiums*. London: HMSO, 1864.

Forbes, Eric G. *Greenwich Observatory*. 3 vols. Vol. 1: Origins and Early History (1675-1835). London: Taylor & Francis, 1975.

Fowler, Hugh. 'Biographical Notice of the late Mr. Thomas Fowler, of Torrington with some account of his inventions'. *Transactions of the Devonshire Association for the Advancement of Science, Literature and Arts* 7 (1875): 171–78.

Franksen, Ole Immanuel. Mr. *Babbage's Secret: The Tale of a Cypher – and APL*. Birkerød: Strandberg, 1984.

'Babbage and cryptography. Or, the mystery if Admiral Beaufort's cipher'. *Mathematics and Computers in Simulation* 35 (1993): 327–67.

Gash, Norman. *Sir Robert Peel: The Life of Sir Robert Peel after 1830*. 2 ed. New York: Longman, 1986.

Glass, D.V. *Numbering the People: the Eighteenth-Century Population Controversy and the Development of Census and Vital Statistics in London*. London: Gordon & Cremonesi, 1973.

Grant, George B. 'On a New Difference Engine'. *American Journal of Science and Arts* II, August (1871): 1–5.

Grattan-Guinness, Ivor. 'Work for Hairdressers: The production of de Prony's logarithmic and trigonometric tables'. *Annals of the History of Computing* 12, no. 3 (1990): 177–85.

'Charles Babbage as an Algorithmic Thinker'. *IEEE Annals of the History of Computing* 14, no. 3 (1992): 34–48.

Hall, Marie Boas. *All Scientists Now: The Royal Society in the nineteenth century*. Cambridge: Cambridge University Press, 1984.

'The "Distinguished Man of Science"'. In *John Herschel 1792-1871: A Bicentennial Commemoration*, edited by D.G. Kinge-Hele. London: The Royal Society, 1992.

Herken, Rolf, ed. *The Universal Turing Machine: A Half-Century Survey*. Oxford: Oxford University Press, 1988.

Herschel, John F. W. 'Report of the Royal Society Babbage Engine Committee': Royal Society, 1829.

Hind, John Russell. *Interpolation Tables used in the Nautical Almanac Office*. London, 1857.

Hoskin, Michael. 'Astronomers at War: South v. Sheepshanks'. *Journal for the History of Astronomy* 20, no. 62 (1989): 175–212.

Howse, Derek. *Greenwich Observatory*. 3 vols. Vol. 3: The Buildings and Instruments. London: Taylor & Francis, 1975.

'The Astronomers Royal and the Problem of Longitude'. *The Antiquarian Horological Society* 21, no. 1 (1993): 43–51.

Huskey, Velma R., and Harry D. Huskey. 'Lady Lovelace and Charles Babbage'. *Annals of the History of Computing* 2, no. 4 (1980): 299–329.

Hyman, Anthony. *Charles Babbage: Pioneer of the Computer*. Oxford: Oxford University Press, 1984.

'Babbage Studies'. *Annals of the History of Computing* 11, no. 3 (1989): 226–7.

Jennings, Humphrey. *Pandæmonium: The Coming of the Machine As Seen by Contemporary Observers*. Edited by Mary-Lou Jennings and Charles Madge. London: Papermac, 1995.

Julius, George. 'Mechanical aids to calculation': Institution of Engineers, Australia, 1920.

Keeler, C. Richard. 'Evolution of the British Opthalmoscope'. *Documenta Ophthalmologica* 94 (1997): 139–50.

King-Hele, D.G., (ed) *John Herschel 1792-1871: A Bicentennial Commemoration*. London: The Royal Society, 1992.

Lardner, Dionysus. 'Babbage's calculating engine'. *Edinburgh Review* 59 (1834): 263–327.

Lindgren, Michael. *Glory and Failure: The Difference Engines of Johann Müller, Charles Babbage and Georg and Edvard Scheutz*. Tran. by McKay, Craig G., *Stockholm Papers in History and Philosophy of Technology*: Kristianstads Boktryckeri, 1987. 2 ed. Cambridge, Massachusetts: MIT Press, 1990.

Lovelace, Ada A. 'Sketch of the Analytical Engine'. *Scientific Memoirs* 3 (1843): 666–731.

MacLeod, Roy. *Public Science and Public Policy in Victorian England*. Variorum Collected Studies Series. Aldershot: Variorum, 1996.

McConnell, Anita. 'Astronomers at War: the viewpoint of Troughton & Simms'. *Journal for the History of Astronomy* 25, no. 3:80 (1994): 219–35.

Instrument Makers to the World: A History of Cooke, Troughton & Simms. York: William Sessions, 1992.

Meadows, A. J. *Greenwich Observatory*. 3 vols. Vol. 2: Recent History (1836-1975). London: Taylor & Francis, 1975.

Menabrea, Luigi Frederico. 'Notions sur la machine analyticque de M. Charles Babbage'. *Bibliothèque universelle de Genève* 41 (1842): 352–76.

Merzbach, Uta C. 'Georg Scheutz and the First Printing Calculator': Smithsonian Institution Press, 1977.

Metropolis, N., and J. Worlton. 'A Trilogy of Errors in the History of Computing'. *Annals of the History of Computing* 2, no. 1 (1980): 49–59.

Moore, Doris Langley. *Ada: Countess of Lovelace: Byron's Legitimate Daughter*. London: John Murray, 1977.

Morrell, Jack, and Arnold Thackray. *Gentlemen of Science: Early*

Years of the British Association for the Advancement of Science. Oxford: Clarendon Press, 1981.

Moseley, Maboth. *Irascible Genius: A Life of Charles Babbage, Inventor.* London: Hutchinson, 1964.

Randell, Brian. 'From Analytical Engine to Electronic Digital Computer: The Contributions of Ludgate, Torres, and Bush'. *Annals of the History of Computing* 4, no. 4 October (1982): 327–41.

'Ludgate's analytical machine of 1909'. *The Computer Journal* 14, no. 3 (1971): 313–25.

Rosenberg, Nathan, (ed) *The American System of Manufactures: The Report of the Committee on the Machinery of the United States 1855 and the Special Reports of George Wallis and Joseph Whitworth 1854.* Edinburgh: Edinburgh University Press, 1969.

Schaffer, Simon. 'Babbage's Dancer and the Impressarios of Mechanism'. In *Cultural Babbage: Technology, Time and Invention*, ed. Francis Spufford and Jenny Uglow. London: Faber and Faber, 1996.

Scheutz, Georg, and Scheutz, Edvard. *Specimens of Tables, Calculated, Stereomoulded, and Printed by Machinery.* London: Longman, 1857.

Singh, Simon. *The Code Book: The Science of Secrecy from Ancient Egypt to Quantum Cryptography.* London: Fourth Estate, 1999.

Smiles, Samuel. *Industrial Biography: Iron Workers and Tool Makers.* London: John Murray, 1876.

Sobel, Dava. *Longitude.* London: Fourth Estate, 1996.

Stein, Dorothy. *Ada: A Life and Legacy.* Cambridge, Massachusetts: MIT Press, 1985.

Stokes, George G. 'Report of a Committee appointed by the Council to Examine the Calculating Machine of M. Scheutz'. *Proceedings of the Royal Society* VII, no. 15, January 21 (1855): 499–509.

Swade, Doron. *Charles Babbage and his Calculating Engines.* London: Science Museum, 1991.

'Building Babbage's dream machine'. *New Scientist* no. 1775, 29, June (1991): 37–9.

'The World Reduced to Number'. *ISIS* 82 (1991): 532–36.

'Redeeming Charles Babbage's Mechanical Computer'. *Scientific American*, February 1993: 86-91.

'Charles Babbage's Difference Engine No. 2: Technical Description': Science Museum Papers in the History of Technology, 1996.

'"It will not slice a pineapple": Babbage, Miracles and Machines'. In *Cultural Babbage: Technology, Time and Invention*, ed. Francis Spufford and Jenny Uglow, 34–52. London, Boston: Faber and Faber, 1996.

Toole, Betty A., (ed) *Ada, the Enchantress of Numbers: A Selection from the Letters of Lord Byron's Daughter and Her Description of the First Computer*. Mill Valley, CA: Strawberry Press, 1992.

Toynbee, William, ed. *The Diaries of William Charles Macready 1833-1851*. 2 vols. London: Chapman and Hall, 1912.

Turvey, Peter. 'Sir John Herschel and the abandonment of Charles Babbage's Difference Engine No. 1'. *Notes and Records of the Royal Society of London* 45, no. 2 (1991): 165–76.

Van Sinderen, Alfred, and Michael R. Williams. 'Happy Birthday Mr. Babbage'. *Annals of the History of Computing* 13, no. 2 (1991): 125–39.

Von Neumann, John. 'First Draft of a Report on the EDVAC': Moore School of Electrical Engineering, University of Pennsylvania, 1945.

Wilkes, Maurice V. 'Babbage as a Computer Pioneer': British Computer Society and the Royal Statistical Society, 1971.

'The Design of a Control Unit - Reflections on Reading Babbage's Notebooks'. *Annals of the History of Computing* 3, no. 2 (1981): 116–20.

'Herschel, Peacock, Babbage and the Development of the Cambridge Curriculum'. *Notes and Records of the Royal Society of London*, 44 (1990): 205–19.

'Babbage's Expectations of his Engines'. *Annals of the History of Computing* 13, no. 2 (1991): 141–45.
Williams, Michael R. 'The difference engines'. *The Computer Journal* 19, no. 1 (1976): 82–9.
'The Scientific Library of Charles Babbage'. *Annals of the History of Computing* 3, no. 3 (1981): 235–40.
A History of Computing Technology: Prentice Hall, 1985.
'Joseph Clement: The First Computer Engineer'. *IEEE Annals of the History of Computing* 14, no. 3 (1992): 69–76.
'The "Last Word" on Charles Babbage'. *IEEE Annals of the History of Computing* 20, no. 4 (1998): 10–14.
Woolley, Benjamin. *The Bride of Science: Romance, Reason and Byron's Daughter*. London: Macmillan, 1999.

Archival Sources

British Library Manuscripts Department. (Charles Babbage correspondence).
Cambridge University Library, Royal Greenwich Observatory Archive (George Biddell Airy papers).
Museum of the History of Science, Oxford (Buxton collection).
Public Record Office, Kew (General Register Office papers).
Royal Society, London (John Herschel papers).
Science Museum Library, London (technical archive).
Waseda University Library, Tokyo (Charles Babbage Collection).

Credits
1,2 by kind permission of Dr. Neville F. Babbage; 3, 40 Mary Evans Picture Library; 4, 5, 6, 7, 8, 9, 10, 11, 12, 14, 15, 20, 21, 22, 23, 25, 30, 31, 32, 33, 34, 35, 36, 37, 38, 39, 42 Science Museum/Science & Society Picture Library; 13 by kind permission of the Nordiska Museet Stockholm; 16 by permission of the President and Council of the Royal Society; 17 by courtesy of the National Portrait Gallery, *London*; 18 © Crown copyright: UK Government Art Collection; 19 V&A Picture Library; 24 Sarah Sceats; 26, 27, 28, 29 Doron Swade; 41 by permission of the Syndics of Cambridge University Library.

INDEX

Ackroyd, Peter, 204
Adams, John Couch, 31, 154
Airy, George Biddell, 138–47; achievements and accomplishments, 139, 188; Babbage's attack on in *The Exposition of (1851)*, 186, 188–9; background and career, 138–9; blunder over Neptune discovery, 153–4; character, 139–40; criticism and condemnation of Difference Engine, 37, 38–9, 126, 137, 141–5, 146–7, 186, 188, 202, 205; eye problem, 182; and Scheutz difference engine, 198, 200–2, 206–7, 208–9; and Sheepshanks, 126, 186
Airy, Wilfrid (son), 139–40
Aitken, Howard, 317
Albert, Prince, 108–9, 185, 199
Analytical Engine, 91–133, 155–71, 179, 210–11, 223–4; achievement of, 114; assembly of the Mill by son after death of father, 4, 315; capacity to automatically repeat a sequence of operations a predetermined number of times, 110; collaboration with Ada Lovelace, 155–6, 161–2, 164–5, 166–9; conception of and progression of ideas from Difference Engine, 91–6; decimal system, 111, 112–13; design, 96–7, 98, 106, 210; difficulty in building a modern day, 224; experimenting with methods of making repeated parts, 214; failure to complete, 122;

Analytical Engine – *contd*
features, 114–15; hiring of Jarvis as principal draughtsman, 97–8, 118; motive behind, 117; negotiations with government and meeting with Peel, 131–2, 149–52; output devices, 106; Prince Albert's visit to, 108–9; protective mechanisms, 111–12; reduction of execution time as ruling principle of design, 98–9, 101; restarting on after ten year gap, 210; size and speed, 111, 115; split of into the Mill and Store, 105; starting of work on Small Analytical Engine, 131; technique of successive carry and anticipating carriage design, 99–103, 104–5; translation of and expansion of Menabrea's report by Ada Lovelace, 160–1, 162–3, 165–6, 169, 171; and Turin convention, 130–1, 132, 160; use of Mechanical Notation as design aid, 119–21; use of punched cards to control, 107–8, 109–10, 166; view of by Plana, 133; vision of as a universal algebra machine, 169, 170; working on in isolation by Babbage, 116, 117, 128
Analytical Society, 18
anticipating carriage mechanism, 102–3, 104, 105
arcades, 181
arithmometer, 11, 28, 83
Association for the Advancement of Science, 74–5
Astronomical Society, 10, 25, 32, 38

Babbage, Benjamin (father), 20–1, 21–2, 23, 50, 52–3
Babbage, Benjamin Herschel (son), 22, 172
Babbage, Charles
Mathematical and Scientific Career:
and Analytical Engine *see* Analytical Engine; and Association for the Advancement of Science, 75; and Astronomical Society, 25; awarded Gold Medal from Astronomical Society for Difference Engine invention, 38; bicentennary celebration and exhibition, 287, 293–4, 296–7, 298–9, 303–6, 308–9; capacity for meticulous detail, 51; criticism of Royal Society in *Reflections on the Decline of Science in England*, 61, 62–5, 87; and Difference Engine *see* Difference Engine; elected Fellow of the Royal Society, 25; elected Lucasian Professor of Mathematics at Cambridge and resignation, 56–7, 126; excluded from Great Exhibition, 185, 186; failure to secure recognition for certain inventions, 182–3; and games-playing machines, 179–80; grievance at lack of recognition, 121, 123, 128–9, 138, 161, 190–1; high regard for abroad, 128–9; on importance of role of computation in advancement of science, 129; influence on modern computing debate and allegation of delaying development, 309–10, 313, 316–18; interest in ciphers, 182–3; interest in tables and publication of logarithms table, 9–10, 50–2; inventions designed, 177–83; lectures on

astronomy, 25; 'limitation of technology' thesis as reason for lack of success, 5–6; and mail-delivery system, 181; and Mechanical Notation, 118, 119–22, 202; meeting with Herschel to verify calculations, 9–10, 15, 17, 25; opposition to patents, 183–4; and ophthalmoscope, 182; preoccupation with mechanised communication, 181–2; railway technology contribution, 127–8, 139, 178; Scribbling Books, 5, 92, 93; and Statistical Society of London, 75; studying at Cambridge and disqualification from sitting finals, 18–20, 23, 57; and theory of miracles, 76, 77–80, 108, 126

Personal Life:
adoration of mother, 22, 173, 191; biographical note, 319–20; character, 31, 73; death of wife and other family members, 52–3, 53–4, 55, 64, 103, 173; depression and loneliness, 190, 191; deterioration of health and death, 216–17; dislike of and campaign against street musicians, 211–13, 216–17; domestic life, 172; early rejections in finding a job, 23; and empiricism, 79; finances, 23, 24, 53; hearing disorder, 214; holiday in Paris, 52; ill-health, 59–60; involvement in litigation case between South and Troughton & Simms, 126, 187; last years, 215–16; marriage, 20–1; political activity, 73–4; relationship with father, 20–1, 21–2, 52–3; relationship with sons, 172; religious views, 19–20; social life and parties held, 72–3, 77–8, 126, 172–3; tour of Continent, 54–6

Writings: articles and papers written, 25, 50, 74, 119; book on life assurance, 50; bypassed in writing one of the *Bridgewater Treatises*, 185–6; *Exposition of (1851)*, 186, 188–9, 190–1, 192; *Ninth Bridgewater Treatise*, 80, 126, 186; *On the Economy of Machinery and Manufactures*, 86–7, 185; output, 215; *Passages from the Life of a Philosopher*, 54, 55, 131, 160, 190, 211

Babbage, Charles (son), 53
Babbage, Dugald (son), 54, 72–3, 116, 172
Babbage, Elizabeth (mother), 22, 59, 98, 173, 191
Babbage, Georgiana (daughter), 54, 103
Babbage, Georgiana (wife) (née Whitmore), 20–1, 22, 40, 52–3, 64
Babbage, Henry (son) *see* Prevost, Henry
Babbage, Mary ('Min', wife of Henry), 216
Babbage, Neville (great-great-grandson), 214, 293
Baily, Francis, 310
Baker, Kenneth, 223
Ball, Mike, 227, 233, 245, 247
Barlow, John, 189
Board of Longitude, 36
Boole, George, 84–5
Boolean algebra, 84
Bramah, Joseph, 41
Brenner, Sidney, 260
Bridgewater Treatises, 185–6
Brigand (iron steamer), 14

Bromhead, Edward, 38
Bromley, Dr Allan, 228, 233, 237, 277; on Babbage, 309; finding a solution to design problem, 239–40; proposal to build Babbage Engine, 221–2, 223, 225–6; working out cost of Engine, 252–3
Brown, Neil, 294
Brunel, Isambard Kingdom, 42–3, 127, 128
Brunel, Marc Isambard, 40–1
Bryden, David, 227
Buckland, William, 135, 136, 137
Burton, Gilly, 294
Buxton, Harry Wilmot, 85
Byron, Lady Annabella, 155, 156, 171
Byron, Lord, 117, 155, 156, 161

cadastral tables, 33, 52
calculation: use of printed tables for *see* tables, mathematical
calculators: devising of automatic, 313–14; devising of mechanical, 10–11, 83, 99; and Fowler, 310; introduction of the arithmometer, 11, 28, 33; problems with early, 27–8; *see also* Difference Engine
Caledonian Canal, 135–6
Callet, François, 51
Canadian Broadcasting Corporation, 249
Carlyle, Thomas, 16
Carnot, Lazare, 33
central heating system, 180
Channel 4, 286
ciphers, 182–3
Clement, Joseph, 4, 65–71, 91, 229, 282; allegations of prolonging Difference Engine contract, 68–9, 97; awarded Society of Arts gold and silver medals, 69; background and engineering career, 41–3, 70; character, 42; collaboration with Babbage in making Difference Engine, 43, 44, 47–8, 60–1, 67–8, 285; confrontation with Babbage over bills and stoppage of work, 66–7; demand for compensation when asked to move into new assembly site and Treasury's rejection, 65–6, 66; hiring of by Babbage, 43; large bills demanded for work, 42–3; problems in collaboration with Babbage, 60–1; work on standardisation of screw threads, 46–7, 69
Clerk, Sir George, 135
Cohen, I.B., 317
Collier, Bruce, 168
Colmar, Thomas de, 8, 11, 28
Computer Conservation Society, 286–7
Computer Talk, 250
Computer Weekly, 248
computers: Babbage's influence on, 308, 309–10, 313, 316–18; and 'debugging', 291; movement leading to, 316; separation of Store and Mill in Analytical Engine as fundamental feature of, 105–6; and Von Neumann, 105–6
'computers' (people), 10, 15, 33–4
Comrie, L.J., 313, 314
Cossons, Neil, 234, 235, 259, [261], 271, 272, 302, 304
Crick, Reg, 241, 242, 244, 272, 277, 280, 280–1, 288
Crimean War, 183
Croker, John, 36
Crosse, John, 171

Dalton, John, 121
Dan Leno and the Limehouse Golem (Ackroyd), 204

INDEX

Darwin, Charles, 73
Davison, Peter, 294
Davy, Sir Humphry, 32, 34, 35, 63
De Morgan, Augustus, 128, 157, 167, 310
De Morgan, Sophia, 167
De Prony, Baron Gaspard, 33–4, 52
Deacon, Alfred, 313
'debugging', 291
Decline see *Reflections on the Decline of Science in England*
Derby, Lord, 176
Dickens, Charles, 70–1, 73, 74–5
Difference Engine (No. 1), 4, 9–87, 111; abandoning of project by government after detrimental reports, 140–8; and Ada Lovelace, 159; announcement of invention, 32–3; article on by Lardner in *The Edinburgh Review*, 193, 195–6; assembly of demonstration piece, 65–6, 233; collaboration between Babbage and Clement, 43, 44, 47–8, 60–1, 67–9, 97, 285; constructing and completion of model, 26–7, 32; costs, 54, 65, 67, 136; criticism of and condemnation by Airy, 37, 38–9, 126, 137, 141–5, 146–7, 186, 188, 202, 205; demonstrations of by Babbage to visitors, 80–1, 247; design, 43–4, 45; drawings, 224–5; exclusion from Great Exhibition, 185; exhibition of at London Exhibition (1862), 213; experimental assemblies constructed, 231; failure to complete after breach with Clement, 67–8; fate of components after ending of project, 67; funding from Treasury, 35–9, 58–9, 61, 135–6, 198, 252, 269; gives impetus to notion of 'thinking machine', 85; government discussion over continuation of, 123, 125; mechanism for carrying tens, 28–9, 99, 232, 237; need for production of large number of similar components and problems encountered, 44–6; original stimulus for, 117; and Peel, 35–6, 134–7, 151–3, 160; personal and financial sacrifices made by Babbage, 58; precision as critical factor in failure thesis, 229–30, 306; and principle of 'method of finite differences', 28–30, 113; problem of resetting machine for each new run of calculations and solution, 94–6; reception to model, 34–5; research and planning, 39–40; and Royal Society, 32, 36–8, 59, 136, 141; rumblings of discontent on money spent on with nothing to show, 57–8; size, 48; slow progress, 46, 65; symbolised start of automatic computation era, 83, 84; uniqueness of demonstration piece in technology history, 82–4; used to demonstrate theory of miracles, 78–9, 108
Difference Engine (No.2), 45, 172–7; conception of and motives behind, 173–4; design, 174–5, 223; drawings, 227–8; efficiency, 225; printing apparatus for, 174–6; turn down of support by government, 176
Difference Engine (No.2) (modern version), 2, 221–318; appearance, 255, 300; assembling of, 283–5, 287–94, 297–8, 299–300; building of as vindication and commemoration of Babbage's

Difference Engine – *contd*
work, 225, 226; carry mechanism, 295; as centerpiece of exhibition, 299; construction of trial piece, 231, 232–51, 254; cost, 252–3, 257, 278; deal with Rhoden, 269–72, 277, 278–9; discovery of flaw in mechanism for carrying tens and finding a solution, 237–40, 242, 250; drawings, 224, 225, 227–8, 237, 270, 272–3, 277–8; early proposals to build, 222–3; funding from Science Museum, 245–6, 253, 272; and IBM, 260–1, 262–6; lubrication problem, 273–4; making of parts using modern techniques, 254–5; and media, 248–50, 265, 279, 285–6, 298, 302–3; metals used to build and analysis of metals used in original engine, 274–6; photograph of, 297; and precision, 230–1; printing apparatus, 255–6, 277; problem of debugging and fault detection, 291–2; problems and queries encountered, 236, 272–4, 290–1; production of parts, 277–8, 284–5; proposal to build by Bromley, 221–2, 222–3, 225–6; public unveiling of at exhibition, 303–5; reasons to build over other engines, 223–5, 226; Rhoden goes into receivership, 279–81; and Science Museum changes, 234–5; shape of gear teeth question, 242–4; slow progress and delays, 244–5, 246, 249; sponsorship, 256, 257–8, 259, 260–1, 266–8; technical feasibility of construction, 226–9, 231; test run and problems encountered, 300–2; testing, 288, 297, 299–300; timescale, 233–4, 257, 262; turning handle solution, 288–9; working of, 306–7
difference engine (Scheutz) *see* Scheutz difference engine
Disraeli, Benjamin, 16, 176–7
Donkin, Bryan, 4, 60, 207, 208
Dudley Observatory (New York), 203

Edinburgh Review, The, 193, 195
EDSAC, 92
English Life Tables, 203, 205, 207–8
Exposition of (1851), The, 186, 188–9, 190–1, 192
Exton, Dave, 294

Faraday, Michael, 199
Farr, William, 203–5, 206, 207
Follett, Sir William, 135
Fowler, Hugh, 310–11, 311–12
Fowler, Thomas, 310–12

Galignani's Messenger, 56
Galle, Johann, 154
Galloway, Alexander, 41
gambling dens, 49
games-playing machines, 179–80
Gardner, John, 267–8
'gauge war', 127, 139
General Register Office, 203, 205, 206
Gilbert, Davies, 35–6
Goulburn, Henry, 13, 137–8; correspondence with Airy over Engine, 138, 140–1, 143, 146–7, 205; letter to Babbage abandoning Treasury support for Engine, 147–8, 149
Graham, George, 205, 208
Grant, Barnard, 314
Gravatt, William, 208, 213
Great Exhibition (1851), 184–5, 186
Great Exposition (Paris) (1855), 202
Greig, Woronzow, 73

Harrison, John, 312
Harvard Mark I, 317
Heaviside, Richard, 194
Helmholtz, Hermann von, 182
Herschel, John, 13, 38, 53; astronomical career, 20, 137; awarded baronetcy, 137; death, 216; declinist views, 64; graduation, 20; meeting with Babbage to verify calculations, 9–10, 15, 17, 25; relationship with Babbage, 18, 19, 21, 63, 137–8; report on Difference Engine, 146; seeking of advice from by Goulburn, 137–8; supervision of Difference Engine project during Babbage's absence, 57–8
Hind, J.R., 205
Hollerith, Herman, 315–16
Holloway, Barrie, 278, 280–1, 288
Hyman, Anthony, 194–5, 222–3, 244

IBM, 3, 240, 259–60; and Difference Engine, 260–1, 262–6; sponsorship of Leonardo da Vinci exhibition, 261
ICL, 268
Illustrated London News, 14, 199
Industrial Revolution, 12
Information Age Project (IAP), 258–9, 266, 286
Isaac, Harriet, 54

Jacquard loom, 107–8, 109, 166, 185
Jarvis, Charles Godfrey, 68–9, 97–8, 118
Jephson, Reverend Thomas, 19
John Bull, 67

Kasiski, Friedrich, 183
Kaye, Geoff, 260, 263, 264
King, William *see* Lovelace, Earl of

Lardner, Dionysius, 13, 27, 85, 117, 194–6
Laurence, Samuel, 173
Le Verrier, Urbain, 154, 202
Legendre, Adrien Marie, 33
'legislative': distinction between 'executive' and, 129
Leibniz, Gottfried, 10–11, 83, 84
Leopold II, Duke, 55
Lewis, George, 205
Little Dorrit (Dickens), 70–1
logic: relationship between 'laws of thought' and rules of, 84–5
Lovelace, Ada, Countess of, 155–71; background, 156–7; collaboration with Babbage on Analytical Engine, 155–6, 161–2, 164–5; exaggeration of contribution to Babbage's Engines, 166–9; interest in Babbage's work, 117, 159–60; last months and death, 171; marriage, 157; mathematics education, 156–7; relationship with Babbage, 162–3, 171; self-regard and conviction of own genius, 158–9, 161–2, 163, 168; translation of and expansion of Menabrea's report on Analytical Engine, 160–1, 162–3, 165–6, 169, 171
Lovelace, Earl of (William King), 157
Ludgate, Percy, 314
Lyell, Charles, 73, 76–7

MacCullagh, Professor James, 130
Macready, William, 73, 145
mail delivery system, 181
Margaret, Dame, 231
Martineau, Harriet, 73
mathematical tables *see* tables, mathematical
Maudsley, Henry, 41
Mechanical Notation, 118, 119–22, 202

Mechanics Magazine, 193
Melbourne, Viscount, 125
Menabrea, Luigi, 132–3, 160
Mensdorf, Count, 108
mesmerism, 75
'method of finite differences', 28–30, 113
'microprogram', 97
miracles: Babbage's theory of, 76, 77–80, 108, 126
Mosely, Maboth, 63
Moulton, Lord, 214–15
Müller, Johann Helfrich, 30–1, 83, 293
musicians, street, 211–13, 216–17

Naisbitt, John, 258
Napoleon III, 202
Nature, 265
Nautical Almanac, 13, 208, 313
Nautical Almanac Office, 205, 206
navigation, 12
Neptune, 153
New Scientist, 302
Ninth Bridgewater Treatise, 80, 126, 186
Notation *see* Mechanical Notation

Observer, 250
On the Economy of Machinery and Manufactures, 86–7, 185
ophthalmoscope, 182

Parsons, Eric, 269–70, 279–80
Pascal, Blaise, 10, 83, 84
Passages from the Life of a Philosopher, 54, 55, 131, 160, 190, 211
patents: Babbage's opposition to, 183–4
Peel, Robert: and Analytical Engine 148–9; and Babbage's Difference Machine, 35–6, 134–7, 151–3, 160
Philosophical Transactions of the Royal Society, 119–20
phrenology, 75
Plana, Giovanni, 129, 130, 131, 132, 133
Playfair, Lyon, 81
polynomials, 29
Prandi, Fortunato, 133
Prevost, Henry, 54, 72–3, 121, 173, 202; continuation of father's work, 4, 67, 314–16; and father's death, 216–17; military service in India, 172; relationship with father, 116, 172
Protector Life Assurance Company, 50

Quételet, Adolphe, 117–18

railways: Babbage's contribution to technology, 127–8, 139, 178; 'gauge war', 127, 139
'Rainbow Dance', 178
Record, 57
Reflections on the Decline of Science in England, 61, 62–5, 87
Reform Bill (1832), 73
Reid, John, 269, 271, 279
religion: and science, 75–6
Rennie, George, 60
Rhoden Partners Ltd, 253; contracting of to produce drawings, 240–2; deal with Science Museum over Engine, 269–72, 277, 278–9; goes into receivership, 279–81
Robinson, John (Viscount Goderich), 38, 56, 58
Robinson, Thomas Romney, 144
Roget, Peter Mark, 199
Rosse, Lord, 176
Royal Observatory, 206
Royal Society, 74; Babbage's anger at

not receiving Royal Medal from (1826), 121; Babbage's criticism of in *Reflections on the Decline of Science in England*, 61, 62–5, 87; debarring of women, 157; and Difference Engine, 32, 36–8, 59, 136, 141; founding, 24–5; movement to reform, 61–2, 64; and Scheutz's difference engine, 199, 200

Sardinia, King and Queen of, 55
Scheutz difference engine, 4, 196–203; Airy's view of, 198, 200–2, 206–7, 208–9; building of, 196–9, 207; exhibition of and awarded medal, 202; failure to deliver benefits in producing *English Life Tables*, 207–8; and Farr, 205; funding, 198; government investigation into whether or not to fund a copy of and recommendation to do so by Airy, 205–7; promotional booklet, 202–3; purchase of and cost, 203; reaction to success of by Babbage, 199–200; and Royal Society, 199, 200; success in England, 199
Scheutz, Edvard, 4, 197, 209
Scheutz, Georg, 4; death, 209; interest in Babbage's Difference Engine, 193–4
Schickard, Wilhelm, 10
science: flourishing of, 74–5; fragmenting of into specialisms, 74; lack of recognition of as a career, 24; and religion, 75–6; *see also* Royal Society
Science Museum, 2; archive of Babbage's work, 4–5; changes in, 234–5; funding of Difference Engine, 245–6, 253, 272; and Information Age Project, 259
Scientific American, 248–9, 309
Scientific Memoirs, 160, 163
screw threads, standardised, 46–7, 69, 229, 230
Sheepshanks, Reverend Richard, 126, 186–8, 189
Sheridan, Wendy, 294
ships: signalling to and from, 181–2
shipwrecks, 13–14
slide rules, 11
Smedley, Reverend Edward, 59–60
Smiles, Samuel, 48
Smith, Adam, 33
Somerset, Duke of, 65, 160, 173
South, Sir James, 126, 151, 186–7, 188
Spring-Rice, Thomas, 125
standardised screw thread, 46–7, 69, 229, 230
Statistical Society of London, 75
statistics: belief in as salvation from suffering, 204; growing fashion for, 76
steam engines, 17
Stephenson, Robert, 67, 127
stepped drum, 11
Swedish Academy of Sciences, 198

tables, mathematical: Babbage's interest in and publication of logarithms table, 9–10, 50–2; human fallibility as source of errors, 14–16; importance to navigation and consequences of inaccuracy, 12, 13–14; inaccuracy of and errors within, 10, 12–14, 27; increase in reliance on, 12; need for and importance of, 11–12; process of compiling, 14–16; production of cadastral tables, 33
Talbot, Fox, 73
Tennyson, 77

Thatcher, Margaret, 2, 234
Thermosiphon, 311
Thwaites, John, 183
Ticknor, George, 80
Times, The, 248
Troughton & Simms, 126, 187, 188
Troughton, Edward, 126, 187
Turin convention, 130–1, 132, 160
Turing, Alan, 84, 170
Turvey, Peter, 236–7, 239, 247, 275

Victor Immanuel II, King, 131
Vigenère Cipher, 183
Vlacq, Adriaan, 52
Von Humboldt, Alexander, 131–2
Von Neumann, John, 84, 105–6
Von Vega, Georg, 51, 52

Wellington, Duke of, 58–9, 108, 123, 124
Weston, Dame Margaret, 222, 234, 236
Wharton-Jones, Thomas, 182
Wheatstone, Charles, 160, 199
Whewell, William, 24, 145–6
Whitmore, Wolryche, 56
Whitworth, Joseph, 47, 66, 69, 210, 229
Wiberg, Martin, 293, 313–14
Wilkes, Maurice, 92, 114, 116, 309–10, 312–13
Wittenberg, Gunther, 241
Wollaston, William Hyde, 34–5, 59, 184
Word to the Wise, A (pamphlet), 74
Wright, Michael, 229, 255
Wright, Richard, 54, 216

Young, Dr Thomas, 37–8, 141

Zuse, Konrad, 312